Practical Astronomy

D0113162

Springer
London
Berlin
Heidelberg
New York
Barcelona
Hong Kong
Milan
Paris
Singapore
Tokyo

Other titles in this series

The Observational Amateur Astronomer
Patrick Moore (Ed.)

The Modern Amateur Astronomer
Patrick Moore (Ed.)

Telescopes and Techniques
C. R. Kitchin

Small Astronomical Observatories
Patrick Moore (Ed.)

The Art and Science of CCD Astronomy
David Ratledge (Ed.)

The Observer's Year
Patrick Moore

Seeing Stars
Chris Kitchin and Robert W. Forrest

Photo-guide to the Constellations
Chris Kitchin

The Sun in Eclipse
Michael Maunder and Patrick Moore

Software and Data for Practical Astronomers
David Ratledge

Amateur Telescope Making
Stephen F. Tonkin

Observing Meteors, Comets, Supernovae
and other Transient Phenomena
Neil Bone

Astronomical Equipment for Amateurs
Martin Mobberley

Transit: When Planets Cross the Sun
Michael Maunder and Patrick Moore

Practical Astrophotography

Jeffrey R. Charles

With 71 Figures

Springer

Jeffrey R. Charles
Versacorp, 2454 E Washington Blvd, Pasadena,
CA 91104, USA

Cover illustrations: Background: Star trails, photographed with a 35 mm wide angle lens. Comet photo: Comet Hyakutake. Left inset: M42, the great nebula in Orion. Center inset: The total solar eclipse of 3 November, 1994. Right inset: 26 October, 1984 conjunction of the moon and Venus. Photos by the author.

ISBN 1-85233-023-6 Springer-Verlag London Berlin Heidelberg

British Library Cataloguing in Publication Data
Charles, Jeffrey
    Practical Astrophotography. – (Practical astronomy)
    1. Astronomical photography – Handbooks, manuals, etc.
    I. Title
    522.6'3
ISBN 1852330236

Library of Congress Cataloging-in-Publication Data
Charles, Jeffrey, 1957–
    Practical Astrophotography / Jeffrey Charles.
      p. cm. – (Practical astronomy)
    Includes bibliographical references and index.
    ISBN 1-85233-023-6 (alk. paper)
    1. Astronomical photography–Amateurs' manuals. I. Title.
    II. Series.
QB121.C48 2000
522'.63–dc21   99-053334

Typeset by EXPO Holdings, Malaysia
Printed and bound at the Cromwell Press, Trowbridge, Wiltshire
58/3830-543210 Printed on acid-free paper SPIN 10662969

In Memory of Pierre-Yves Schwaar
1946–2000

Pierre was an expert telescope maker, serious observer, and friend.
The telescopes he made are windows to the heavens through which
many can witness the magnificent wonders of the universe.

The summer Milky Way is impressive from a dark site. The 10 minute exposure at f/2.8 on Ektachrome 200 film was taken with a Minolta SRT 101 camera and 16 mm Rokkor-X fisheye lens. Photo by the author, Jeffrey R. Charles. Unless otherwise noted in the caption, other photos and artwork in this book are by the author.

Preface
Capturing the Wonder of the Sky

The sky is full of wonders. It can be dynamic. It can be a deep blue. It can be filled with interesting clouds. It can show the bright colors of the rainbow. It can herald the approach of the moon's awesome shadow before a total solar eclipse. The night sky in particular is full of nebulae and other exotic objects. These relatively dim objects are always present, but the brightness of the daytime sky hides them from view. At times, the night sky is host to awe inspiring events including lunar eclipses, occultations of planets and stars by the moon, conjunctions, meteor showers, and the approach of impressive comets.

After seeing such marvelous objects and events, it is only natural to want to photograph them. As a result, many people have become involved in the exciting field of astrophotography. Years ago, astrophotography was very difficult due to a lack of good equipment and films. Fortunately, the related technologies matured over time as innovative individuals and small businesses developed new techniques, telescopes, and astrophotographic accessories. Films have continued to improve, resulting in better resolution and shorter exposure requirements. Today, astrophotography is relatively simple, though a few specialized procedures are still admittedly a bit tedious.

You can get into basic astrophotography with simple equipment. There is no need to be a dedicated "amateur astronomer" or wait until you get that big telescope or the latest astrophotography gadget before you go out and start shooting your own celestial photos. You can photograph some celestial objects without any telescope at all. A simple camera and lens can get you started, opening the door to a surprising variety of objects, from lunar phases, to planetary conjunctions, to time exposures of constellations or star trails. You can even use the twilight sky to silhouette terrestrial objects in some of your pictures.

With a few dollars worth of materials, you can build a simple mount which will allow you to manually track the stars with your camera, facilitating long exposures which reveal star clouds, nebulae, and even the fleeting passage of a major comet. If you have a telescope, more objects will be within your reach. A good telescope or long telephoto camera lens as small as 4 cm aperture can provide a decent image of the moon, and a good telescope of only twice this aperture can provide useful images of planets and many other objects.

Whatever your reason for getting into astrophotography, it can be a rewarding experience. Some take astrophotos for research; others do it for the challenge; others do it just for fun; and still others may do it to remember various sights and events. It is important to remember that even a "bad" picture can be better than no picture. Just like other images, practically any astrophoto can bring back memories of the night you took it, the heavenly objects that impressed you most, the telescope you used, the site you were at, or the friends you were with. The night sky is there for all of us to enjoy. In capturing the wonder of the sky, you can capture a memorable part of your life!

The Scope of This Book

This book is written as a useful resource for both photographers and amateur astronomers. Whether you are a photographer who just wants to experiment with shooting celestial objects; a beginning amateur astronomer who wants to take a picture of the moon; a traveler who wants to photograph an eclipse; a student who wants to learn specific techniques; or a seasoned astrophotographer; this book will probably have something that is just right for you.

Chapter 1 is an introduction to photography, covering basic topics like shutter speeds, f/stops, focal length, and how to use a manual camera. Chapter 2 covers basic astrophotography and briefly describes a few types of telescopes and mountings. Chapters 3 and 4 introduce how optics work and point out some of their limitations. Included are a few pointers on how to figure out the focal length and f/ratio of your telescope when using a Barlow lens or other auxiliary optic; a good thing to know if you are to arrive at the correct exposure without a lot trial and error. Chapter 5 deals with combining astrophotography with domestic and foreign travel.

Chapters 6–9 cover numerous effective techniques for different types of astrophotography, including how to photograph events such as conjunctions and eclipses. In general, this group of chapters starts out by showing what you can do with very modest equipment, then progresses to more advanced and specialized techniques. Specific topics include photography from a fixed tripod; methods for briefly tracking stars without a conventional equatorial mount; polar aligning an equatorial mount; all-sky photography; piggyback photography with wide angle, normal, and telephoto lenses; photography of solar system objects through both alt-zimuth and equatorially mounted telescopes; close up solar, lunar, and planetary photography through telescopes which are used with auxiliary optics; off-axis guided deep sky photography; and guided comet photography.

Chapter 10 covers how to get the most out of photos after shooting them, including how to identify and correct certain problems you may encounter when shooting your first astrophotos. Also covered are a few basics about darkroom techniques and digital image processing.

The appendices include an astrophotography exposure guide which may save you the cost of this book in film alone. Also included in the appendices are sample checklists for star parties and more extended expeditions. Throughout the book are many celestial photos which include exposure data. These can be used as an additional reference to assist you in taking your own pictures.

When I started taking astrophotos, I learned how to do it by just going out and shooting pictures rather than reading every book I could find on the subject. In this book, I'm primarily passing along what I learned from my own experience rather than just parroting what I heard or read at one time or another. As a result, some material may be presented in a unique (though effective) way.

This book covers film photography almost exclusively, though astronomical imaging with a camcorder is briefly covered because some readers may already have one. Integrating electronic imaging cameras are not covered; that is a subject worthy of a separate book in the *Practical Astronomy* series by those who specialize in the field.

A few basics about astronomical gadgets are covered in this book because the proper use of equipment is how you get results. Results simply won't happen if your telescope and accessories just sit around in a closet.

With today's cool astro gadgets and fast, sharp films, one can often get excellent results with comparatively little effort. Computer related technology has been applied to the field of astrophotography to such an extent that one can literally take astrophotos in their sleep. It is even likely that a trained monkey could take good astrophotos with today's equipment; however, not all of us can afford to throw enough money at our hobby to get all of the latest optical and electronic astro gadgets.

This book is written with an eye toward those of us who can't just go out and buy any gadget we want. An emphasis is placed on showing that you may be able to shoot good astrophotos with a camera lens or small telescope you might already have or reasonably be able to afford to buy or make. Many of the photos show what level of results can be expected with certain types of affordable equipment.

I have used cameras of various formats and many different types and sizes of telescopes, but most of the photos I selected for this book were taken with a 35 mm camera and a lens or telescope having an aperture of 102 mm or less. This is the size of optics that many amateur astronomers have. An amazing variety of subjects are within the reach of relatively modest optical equipment, and even more subjects can be imaged if we are willing to work for our results. Effort can narrow the gap between the serious amateur astronomer with modest equipment and the casual observer with fancy equipment. If your equipment is better than what I used for the photos in this book, you should get superior results with less effort.

Some may have the preconception that a telescope is necessary for astrophotography. The reality is that a variety of celestial objects can be photographed without any telescope at all. In fact, many widely published photos of eclipses, major comets, and other objects were taken through camera lenses rather than telescopes. A good part of this book is devoted to showing you how to get the most out of various camera lenses. If you have a camera, you may be able to get started in some type of astrophotography; maybe even tonight!

Conventions

This work typically adheres to convention in fields such as astronomy, optics, and the like. I departed from convention in the way a few mathematical equations are expressed; instead I present them in a format that can make it relatively easy to do the math in your head while using your telescope, hopefully without having to resort to a calculator or writing instrument. Expressed units of measure in this book are typically metric, but some metric units may be followed by equivalent imperial units, most often in parentheses.

Most of the conventions used in this book are from related technical fields, while others may have more informal origins. The latter type of convention may not always be consistent. For example; camera lens data is typically specified in terms of focal length and f/ratio, while telescopes are most often specified in terms of aperture and f/ratio. Therefore, a given telephoto camera lens may be specified as a 600 mm f/8, while a telescope having the same optics may be specified as a 75 mm f/8. This book uses these conventions, but in order to minimize confusion, centimeters are typically used when expressing units of aperture, while millimeters are typically used for focal length. Therefore, a 600 mm f/8 camera lens would still be called a 600 mm f/8, but a telescope of the same aperture and focal length would be called a 7.5 cm f/8.

Acknowledgments

I would like to thank fellow astrophotographers who have conceived and implemented improvements in the field of astrophotography over the years; particularly those who contributed photos and provided information about their techniques. I also thank my parents, J. Randall and Wilma D. Charles, for their support during the time I took from my business to write this book; and, Art Montes De Oca for allowing me to use his scanning and CD writing equipment for this project. Most of all, I thank the Lord for creating the celestial wonders that are the subject of interest for astrophotographers all over the world. Finally, I am grateful to John Watson, Managing Director of Springer, UK, and Nick Wilson, Senior Production Controller, for supporting the content of this book and allowing me to give this project the time it deserves.

About the Photo Contributors and the Author

David L. Charles (the author's brother) contributed the photos for Figure 6.7 and the lower right of Figure 9.1, as well as information about his enhanced "Barn Door" mount. David is a computer consultant in Loveland, Colorado. He is also the founder of Colorado Computer Museum (www.trailingedge.org), a nonprofit organization which is dedicated to the preservation and interactive exhibition of information (computer) technology of all eras.

Jeffrey R. Charles (the author) provided most of the photos for this book. Jeff is an engineer at the Jet Propulsion Laboratory in Pasadena, California, and is also the founder and owner of Versacorp, a company which has supplied unique multiple function instrumentation for amateur astrophotography since 1984. Many products marketed by Versacorp or authorized licencees are based on the wide variety of gadgets Jeff has invented over the years for his own amateur astrophotography. More recently, the emphasis of

Versacorp has shifted to licensing authorized vendors to manufacture and market the patented optical systems Jeff has invented, including astronomical instrumentation and catadioptric optics for one shot immersive panoramic imaging, virtual reality projection systems, and other applications. Jeff is a co-founder of the Verde Valley Astronomy Club in Arizona and is an amateur astronomer and total solar eclipse chaser.

Germán Morales Chávez contributed the photo in Figure 6.2. He is the director of Astronomia Sigma Octante (ASO) a center for astronomical research in Cochabamba, Bolivia, which has been involved in observation and public education for more than 20 years. The author first met Germán when visiting Bolivia for the 3 November, 1994 total solar eclipse. Even at that time, Germán had personally documented far more astronomical observations than any person the author had ever met.

Richard Payne contributed the photos in the lower right of Figure 6.7 and the upper right of Figure 9.1. Richard is an avid amateur astronomer who lives in Avondale, Arizona. He formerly owned Metal Craft, a metal working company which manufactured weight machines, plus a unique observer's chair which was well suited for use with a fork mounted Cassegrain telescope.

Pierre-y Schwaar of Phoenix, Arizona, contributed the images in the right of Figure 2.2 and the lower right of Figure 8.2. For decades, Pierre has been implementing innovative commercial and amateur astro gadget projects which include a wide range of telescopes, mountings, guiding systems, and just about anything else that an astrophotographer would need. Pierre is well known for his excellent primary mirror making and for inventing the legendary "Bigfoot™" telescope mount and the adjustable "Piccadilly" partial aperture guider for Newtonian telescopes.

Carina Software's Voyager 2 program was used to obtain stellar coordinates for the chart in Figure 7.2 and to generate the right images in Figures 10.7 and 10.8.

Contents

Chapter 1

Photography Basics

With today's automatic cameras, basic photography is literally a snap. You don't need to know about f/stops, shutter speeds, or other photographic terms; an automatic camera does virtually everything for you in ordinary daylight or indoor settings.

Astrophotography involves taking pictures under less conventional conditions, so it can require a bit more effort. More importantly, it becomes particularly useful to know how your final image will be influenced by the use of a given film, f/stop, or shutter speed. In addition, astrophotography often requires the use of shutter speeds that are outside the range provided by most automatic cameras. Therefore, a camera having manual settings is best for many types of astrophotography.

This chapter covers a few basics about photography and manual cameras. We will start with basic information about films, cameras, and photographic nomenclature, then look at some ways that various film speeds and camera settings are interrelated. Those already familiar with these subjects may want to skip ahead a few pages.

1.1 Cameras

A camera is the most basic piece of equipment you are likely to use in astrophotography. Some cameras work better than others for certain types of photography, but the ones that work best for astrophotography do not necessarily have to be expensive.

Some cameras are completely manual. Others have many automatic features. Completely manual 35 m cameras have manual film loading, manual film advance, manual exposure settings, manual focus, and a manual film rewind crank. Some manual cameras have built-in light meters and others don't. Some have a mechanical shutter speed control mechanism while others have an electronically controlled shutter which may allow for automatic exposure control. Of these, a mechanically controlled shutter has the advantage of being operable without a battery. You can usually identify a mechanically controlled shutter by its sound. When operated at the slowest timed speed, most mechanical shutters make a buzzing sound while they are open.

Interchangeable lenses are another important feature. This will allow you to switch lenses and even mount your camera directly to most good telescopes with an appropriate adapter. Manual cameras are becoming more and more difficult to get new. This often means that you have to look for a used one. A used manual camera can be a lot less expensive than a new automatic one, but it is important to be sure that a used camera works properly (or that the seller will guarantee it works) before you buy it.

If you have never used anything but a modern "point and shoot" camera and are not familiar with manual camera settings, never fear. Using a manual camera is not difficult, and knowing a few basics about camera settings will make it even easier for you to use one and improvise more effectively. It is important to remember that up until the 1970s, most cameras had only manual settings – and people took good pictures with them!

Figure 1.1 is a front view of a manual 35 mm SLR camera body. At center is the lens mount, encircled by the light meter coupling ring. The lens mount index dot is just above the meter coupling ring. The self-timer lever is on the front of the camera, left of the lens mount. Closer to the lens mount are the depth of field preview button (to bottom left of the lens mount) and the mirror lockup control (upper left). The lens release button is just to the upper right of the lens mount, and jacks for a flash unit are toward the right. Visible inside the lens mount are the reflex mirror (center) and the auto-diaphragm control lever (bottom). The auto-diaphragm mechanism allows the aperture of a compatible lens to remain open at all times except the instant the picture is taken, facilitating the brightest possible viewfinder image.

Figure 1.1. Front view of a typical manual single lens reflex (SLR) camera body. An SLR camera is excellent for astrophotography because it allows you to see the actual image formed by a lens (or your telescope) in the viewfinder. This is accomplished by means of a reflex mirror which reflects the image formed by a lens (or your telescope) to the camera's focusing screen. The reflex mirror flips up when a picture is taken.

A top view of the same camera (Figure 1.2) shows it with a normal lens attached. Near the front of the lens is the focus ring and distance scale. Just behind the focus ring is a depth of field scale which also includes a diamond shaped focus index mark. Some lenses have the letter "R" beside the focus index. This is for use with infrared film. (To use the infrared mark, focus the lens in the normal way, then manually turn the focus ring until the distance by the normal index mark is moved to the R mark.) The lens' mount index dot is to the left of the distance scale. Closest to the camera is the aperture (or f/stop) ring and its meter coupling tab.

On top of the camera (from left to right) is the combined rewind crank and back latch release (35 mm film must be rewound before you open the camera back); flash shoe; focal plane indicator (the small circle with a line through it); combined shutter speed and ASA (now called ISO) film speed dial. Just right of this is the

Figure 1.2. Top view of the same camera, also showing a normal lens.

Figure 1.3. Lower rear view the same camera, with its back open.

combined film advance and shutter wind lever (the shutter must be manually wound on a camera not having a motor drive); the shutter release button (at the center of the film advance lever hub); and the exposure counter. On most manual cameras, a cable release can be attached to the center of the shutter release button.

The lower rear view of the same camera (Figure 1.3) shows it with its back open. At top center is the camera's eyepiece. Inside (from left to right) are the back latch mechanism; film cartridge cavity and rewind fork; film guide; focal plane shutter; film advance sprocket, and film advance reel. On the back is a pressure plate which keeps the film flat. On the bottom (from left to right) are the light meter switch; a $\frac{1}{4}$ inch, 20 tpi tripod socket; battery holder; and rewind release button. Not all manual cameras have features such as a built-in light meter, mirror lockup, depth of field preview, or bayonet lens mount. Further, various controls are not necessarily in the same place on all cameras.

If you have a general purpose manual SLR camera such as the one shown above, you will most likely be able to use it for both astrophotography and pictures of other subjects such as your telescope setup, observing site, friends, etc. You can also use it with whatever lenses you may have for piggyback photography of the sky. A further benefit of a manual camera is that the viewfinder is usually free of light-emitting indicators. A few automatic cameras have indicators which remain on even during an exposure. If this occurs, the indicators can fog your film during a long exposure.

If you want to use a camera on your telescope, it should accept interchangeable lenses rather than having a fixed lens which merely accepts front attachments. Some types of astrophotography are possible through a telescope with a fixed lens camera, but you will have a lot more flexibility if your camera body can be attached directly to a telescope. This in effect will permit you to use your telescope as a telephoto or super telephoto lens. Fortunately, most manual SLR cameras (and even many automatic SLR cameras) accept interchangeable lenses. Some older cameras accept screw mount lenses which thread into the camera body, but the majority accept bayonet lenses which are easier to attach.

The type of lens mount your camera has will depend on the brand, and in some cases, the era of manufacture. Time of manufacture is relevant for camera brands such as Pentax, since early Pentax cameras had a 42 mm thread mount and later ones have a bayonet mount. Other SLR camera brands such as Canon and Minolta had one type of bayonet lens mount for their early cameras, but a new mount design was used when they introduced their auto focus cameras. Nikon has been thoughtful enough to retain the same basic lens mount for many decades, through there were some notable changes in the light meter coupling ring back in the 1970's.

To attach a 35 mm camera body to a telescope, you will need a "T-ring" (sometimes called a T2 mount or T–adapter) that is made specifically for your brand camera. A T-ring has a "T-thread" on the front and a mount like that of one of your camera lenses on the back. The T-thread (or T-mount) system was developed as a standard by independent lens manufactures so they would have to make only a single model lens of a given focal length and f/ratio. This lens could then be adapted to a particular camera with the appropriate T-ring. Since T-thread offers a standardized interface thread size (42 mm diameter × 0.75 mm pitch) and a fixed adapter flange to focal surface distance (55 mm), it is also useful for adapting optics such as telescopes and microscopes to various 35 mm camera bodies

Many telescopes do not include a T-thread camera interface as standard equipment. In such cases, a separate T-adapter must be acquired. This adapter attaches to the telescope and has a male T-thread on the back. For clarity, this book refers to the adapter which attaches to a telescope and has a male T-thread on the back as a "T-adapter". The term "T-ring" is used for the adapter which has a female T-thread on the front and the camera lens mount on the back.

When using an SLR camera on a telescope, you may notice that the very top of the viewfinder image is not fully illuminated. This is because the reflex mirror in some SLR cameras is not quite long enough to provide full illumination of the entire focusing screen with the image from a telescope. Fortunately, such reduced illumination of the focusing screen will not affect your photograph.

The standard focusing screen in a typical SLR camera is optimized for fast f/ratio lenses. It will usually work for photography through a telescope, but it will tend to look grainy and any central focusing enhancements such as the micro prism or split image circle may black out when used with optics slower than about f/5.6. This will not affect your picture, but it may require you to use the part of the screen just to the side of any blacked out central zone for focusing.

Some 35 mm SLR cameras accept interchangeable viewfinders and focusing screens. Most of these accept right-angle viewfinders which are very useful for astrophotography with Cassegrain and refracting telescopes. If the right-angle viewfinder offers direct viewing of the focusing screen (as opposed to being an attachment for the camera eyepiece) the image will usually be brighter and easier to see. Other accessories include a critical magnifier which offers a magnified view of at least the central part of the focusing screen.

If you want a critical focusing magnifier and one is not offered for your camera, it may be possible to adapt an attachment from another manufacturer to fit; or, you may be able to use an appropriate monocular (such as one 8 × 20 unit made by Specwell) which can focus down to a distance of about half a meter. It must focus this close because the eyepiece in a typical SLR camera viewfinder is designed to cause your eye to focus at about that distance. Other focusing techniques depend on using a separate focusing attachment such as the Spectra Sure Sharp™ which has a multiple knife edge (or Ronchi grating) which is temporarily used in place of the camera. Focusing attachments like the Versacorp MicroStar™ offer the convenience of use in the top of a suitable flip mirror unit.

Other types of manual cameras can be used for astrophotography, but many of these are either rangefinder cameras or special purpose gadgets which lack a built-in viewfinder and must be used with a previewer, flip mirror, or other accessory. Special purpose cameras are usually well suited to certain types of astrophotography, but they may not be very useful for day to day pictures of other subjects. They also represent an additional expense for anyone who may already have a suitable manual 35 mm SLR camera body.

1.2 Exposure Time

Exposure time is an important consideration for any type of photography. If you are photographing fast action and want a sharp image, the exposure must be so short that neither the subject or camera will move much while the shutter is open. If the subject moves too much during the exposure, it will appear to be blurred in the image, regardless of how well the focus is adjusted. Unfortunately, the luxury of a fast shutter speed is only possible under certain conditions. In the case of very dim subject matter, the required exposure time can be so long that some people may actually purchase special large aperture lenses or other equipment that will allow the use of a shorter exposure time.

On most manual cameras, exposure setting is controlled by a shutter speed dial having a series of numbers, each corresponding to a particular fraction of a second. Since a shutter is what typically controls the exposure time, the terms "shutter speed" and "exposure time" are used interchangeably. Numbers on a typical shutter speed dial may include a series of numbers such as 1, 2, 4, 8, 15, 30, 60, 125, 250, 500, and 1000, corresponding to 1 second, $\frac{1}{2}$ second, $\frac{1}{4}$ second, and so on. Each progressively faster shutter speed is roughly half the duration of that which preceded it, and it is easy to see that the numbers used are approximations for powers of two: 2, 4, 8, 16, 32, 64, 128, and so on. Therefore, each shutter speed setting changes the exposure time by roughly a factor of two, or two to the first power. Clicking the shutter speed dial two settings changes the exposure by a factor of four, or two to the second power, and so on.

In astrophotography, exposure times can exceed 1 second by a substantial margin, and the numbers used to represent these longer times can also be expressed as approximations for powers of two: 1 second, 2 seconds, 4 seconds, 8 seconds, 15 seconds, 30 seconds, 60 seconds (or 1 minute), 2 minutes, 4 minutes, etc. These approximations for powers of two conveniently allow for easy conversion between seconds, minutes, and hours. One minute is roughly equivalent to two to the sixth power seconds, or the equivalent of six "shutter speed settings" from a one second exposure time; and one hour is the equivalent of six shutter speed settings from one minute.

1.3 f/stops

The f/stop of an optical system corresponds to the ratio of its focal length to the diameter of its aperture. This is commonly referred to as the focal ratio, or f/ratio, as can be seen from the way the focal length and maximum focal ratio are usually indicated on a camera lens. For instance, a normal lens on a 35 mm camera may have nomenclature which reads: f = 50 mm 1:2, meaning that the focal length is 50 mm and the effective diameter of the aperture is $\frac{1}{2}$ as large as its focal length, or that is an f/2 lens. If you want to know the effective physical aperture size of the lens, simply divide the focal length by the f/number: 50 mm focal length/2 = 25 mm aperture.

The f/stop scale on a typical normal camera lens may include a series of numbers such as 1.4, 2, 2.8, 4, 5.6, 8, 11, and 16. Since the numbers represent the ratio of the focal length to the aperture, it can be seen that the f/1.4 setting would be the largest. This may seem a bit reversed – a smaller f/number results in a larger lens opening and vice versa; however, when you envision light entering through the lens aperture, it is easy to grasp the concept: at a given distance from the film, a larger aperture will result in a brighter image.

You can also envision the concept of f/stops from the vantage point of the film: from the film surface, a faster f/ratio will look like a circle of light having a larger angular size. This will let more light get to the film. The concept is similar to the difference that enlarging a window would make on the daytime brightness of a particular room (a large picture window obviously lets in more total light than a small window), except that camera optics bend or reflect light to form images while the window does not. A brighter image allows a faster shutter speed to be used for a given exposure value, so a lens having a large aperture compared to its focal length lens is often called a "fast" lens. The term "exposure value" is used here to represent the total amount of light admitted to the film when the picture is taken.

It makes sense to use f/ratio designations rather than units of absolute aperture diameter because the f/ratio is what determines how bright the image of a typical subject will be. For example, the image from a 50 mm f/2 lens will be just as bright as the image from a 100 mm f/2 lens even though the aperture of the 100 mm lens has twice the diameter and four times the area. Images of equal brightness result because the image scale of the 100 mm f/2 lens is twice as large as that of a 50 mm f/2 lens, causing light from a given part of the subject to be spread over four times as much area.

This situation would be different if both lenses had the same measured aperture diameter. For instance, if both lenses have an aperture of 25 mm, the 50 mm lens would still have an f/ratio of f/2, but the f/ratio of the 100 mm lens would become f/4. The relative difference in image scale between the two lenses would still be the same, but the image from the 100 mm lens would now be 4 times dimmer than that from the 50 mm lens.

One exception to f/ratio alone affecting image brightness relates to imaging stars, since they are effectively point light sources. Here, the actual aperture area becomes more important than f/ratio in determining how bright a star can be imaged during a given exposure time. A star is effectively imaged as a dot with any good lens, and its light does not become spread out over a larger area as the focal length is increased. This is an oversimplification, but it gets the idea across. The subjects of aperture, f/ratio, and stellar image size are addressed further in Chapter 4.

It may initially seem as though there is no rhyme or reason to the f/numbers commonly used, but the numbers really have a very elegant relationship to each other. Each f/number is about 1.4× larger than the one preceding it. Why 1.4×? Because it approximates the square root of two, or about 1.4142. When a surface (in this case the lens aperture) is scaled by the square root of two, its *area* is doubled. Therefore, each successively larger f/stop setting (with the correspondingly smaller f/number) has twice the effective area of the one preceding it, so just as with shutter speed settings, each f/stop setting changes the exposure value by a factor of two! Turning the f/stop scale two settings changes the aperture area and exposure value by a factor of four; turning it three settings changes the value by a factor of eight, and so on.

Stopping a lens down to a smaller aperture is useful for a variety of reasons. Most important is the ability to control the exposure value. In bright daylight conditions, a camera may not have a short enough shutter speed to provide the correct exposure without stopping down the lens. Stopping down a lens provides another degree of freedom in controlling exposure. Stopping down a lens will also increase its depth of field, making subjects at varying distances appear relatively sharp in the same picture. A further advantage can be to improve optical performance, since many fast lenses are not particularly sharp at full aperture.

1.4 Interrelation of Exposure Time and f/ratio

As we have seen, each successive setting on either the shutter speed dial or f/stop ring will change the exposure value by a factor of two. This means that we can easily change the shutter speed and still maintain the same exposure value if we change the f/stop by the same number of units in the correct direction. For example, if the lens's f/stop is opened one stop to make the image twice as bright, only half the exposure time will be required to get the same exposure value that would have been provided by the previous settings. The adjustable dial on a light meter shows this relationship very well, but since many of you may not have a light meter, a few sample exposure scales are shown below instead. Each example shows different combinations of shutter speeds and f/ratios which produce the same exposure value.

All of the exposure settings in the first example will typically provide the same correct exposure value for normal daylight with ISO 100 film (shown exposure times represent fractions of a second unless followed by "s", "m", or "hr"):

Shutter speed:	1000	500	250	125	60
f/stop:		4	5.6	8	11	16

The next scale is for an average cloudy day. Note that the new exposure value is accommodated by simply sliding the shutter speed scale over the f/stop scale.

Shutter speed:	125	60	30	15	8
f/stop:	4	5.6	8	11	16

The final scale is for deep sky astrophotography of a dim deep sky object from a dark site with ISO 800 film. It is here that one can appreciate the advantage of a fast f/ratio!

Shutter speed:	15 m	30 m	1 hr	2 hrs	4 hrs
f/stop:	4	5.6	8	11	16

Below is an extended list of shutter speeds which includes the full range of exposure times that you may normally encounter:

4000 2000 1000 500 250 125
60   30   15   8   4   2
1s   2s   4s   8s   15s 30s
1m   2m   4m   8m   15m 30m
1hr 2hr 4hr 8hr

Here is an extended f/stop scale:

1   1.4 2   2.8 4   5.6
8   11   16   22   32   45
64 90   128 180 256 360
512

You can use these scales to estimate the exposure time required when you change your f/ratio or vice versa. For example, let's say you take an image of the first quarter moon with your telescope and your best exposure was 1/60 second at f/11 on ISO 100 film. Now, say you want to get a bigger image of the moon by using your 2× teleconverter on the same telescope. Since a teleconverter doubles the focal length, the effective f/ratio of your telescope will be reduced from f/11 to f/22. (This change in f/ratio results from using the same aperture diameter to provide twice as much image scale.) To calculate your new exposure time, you can just use segments of the above scales, starting with the exposure time you used for f/11:

Shutter speed:	60	30	15
f/stop:	11	16	22

Therefore, your new exposure is 1/15 second at f/22.
You can obviously use f/ratios between the discrete steps shown. Some camera lenses and telescopes have

maximum f/ratios such as f/2.5, f/4.5, or f/10. For long exposures in particular it is not unusual to also use exposure times which fall between the exact values shown in the table.

1.5 Photographic Films

Film is the medium on which your astrophotos will be captured. You can get black and white film, color film, and specialized film which is most sensitive to a given part of the spectrum. With color film in particular, you can get film for transparencies (slides) or negatives. In many cases, film will be the limiting factor for resolution and the range of subjects you will be able to photograph with a given set of optics. Fortunately, a wide variety of films is available, each with its own unique set of characteristics.

Transparency film is usually the best for your first test shots, since it will give you more or less absolute results. In other words, an underexposed transparency will look too dark and an overexposed one will look too light, so you can instantly tell whether or not a picture is exposed correctly. Results from negative film may be a bit more difficult to quantify because photo labs can (and usually do) compensate for moderate errors in exposure by printing an underexposed photo lighter than usual, or printing an overexposed one darker than usual.

For astrophotography, the subjects you shoot and how you intend to display your images will have a lot to do with which basic type of film you should use. If your final image will be black and white, then black and white film of normal or specialized spectral sensitivity is a good choice. (Films of specialized spectral sensitivity may be hard to find in some areas.) If you want to emphasize subjects having certain colors (such as the red in an emission nebulae for example), you can use an appropriate filter (typically one that transmits the same color as your subject) to do it on black and white film.

An original color image does not always make a good black and white one without some extra effort, but there are some circumstances in which color film can be used to advantage when the final image will be black and white. This is particularly true if you have a good image processing computer program with which you can independently adjust each of the primary colors in an image to emphasize features within a certain color range, then convert the file to monochrome.

Color slide film is a good choice if the resulting image will be reproduced in a printed publication or presented with a projector. A backlit or projected transparency provides a high contrast display, so original slides of the night sky can look really good.

If your primary goal is get good prints for your photo album, color negatives may be a better and more economical choice. Color negatives usually capture more dynamic range (i.e. a wider range of relative brightness) *in the original subject* than slide film, though a typical machine print from a negative is not capable of showing all of this range. You can make transparencies or dodged prints from a negative, but this obviously involves more time and expense.

A print will typically have less dynamic range than your original film. The reduced dynamic range of a print results from light having to pass through the print emulsion twice. First, the light must pass down through the emulsion, then it must reflect off the white backing, and back up through the emulsion. Accordingly, the print emulsion can be only half as dense as that of a good transparency. In addition, the white backing on the print is not 100 percent reflective. You can see why a print's emulsion must be less dense by placing a properly exposed transparency on a piece of white paper and observing how dark it looks when light has to pass through its emulsion twice.

There are times when both the film market and the demands of astrophotography will determine what type of film you use. Sometimes, a manufacturer will come out with a really fast and efficient slide film which leaves all of the negative films in the dust. At other times, the reverse situation may be true. Some astrophotographers have used a hybrid approach of shooting a color slide film, then having it developed in color negative chemistry. This can be fine if you do your own film processing, but it is not always a welcome practice at commercial photo labs because some slide films can prematurely exhaust color negative developer.

1.5.1 Film Sensitivity

For many types of astrophotography, film sensitivity is of paramount importance because it will ultimately determine the exposure time required to get a good image. For some brighter subjects, you can use less sensitive films which are capable of providing the highest resolution.

Photographic film is rated according to its sensitivity to light, and the rating is typically indicated by an ISO or EI number. (The ISO or EI number for a given film is equivalent to the older ASA number.) Modern films have ISO numbers that range from about 25 to 3200, and a high ISO number indicates that a film is more sensitive to light than a film with a lower ISO number.

In photographic jargon, the ISO number is informally referred to as the film "speed". This term is no doubt used because a film having a high ISO number will allow the use of a faster shutter speed than would a film with a lower ISO number. Film with a low ISO number is often said to be "slow", and film with a high ISO number is said to be "fast". There are trades to consider when using fast film. Fast film tends to have a larger grain structure and capture a lower dynamic range (i.e. it typically captures detail in a lower range of relative brightness) than a slower film.

The correspondence between the ISO number and film speed is linear, so quantifying film sensitivity is very straightforward. If the ISO number of a given film is twice that of the ISO number for another film, then the film with the higher ISO number is twice as sensitive to light. If the ISO number is four times larger, then the film is four times more sensitive, and so on. Accordingly, if a "fast" film is four times more sensitive to light than a given "slow" film, it will only require $\frac{1}{4}$ as much exposure as the "slow" film.

This linear film speed relationship is true in theory, but when it comes to long exposure times, many films can depart substantially from a linear sensitivity, becoming less efficient (or effectively "slower") as the exposure time is increased. This variation is called "reciprocity failure".

1.5.2 Reciprocity Failure in Film

The vast majority of photographers shoot their photos in daylight or utilize flash units to illuminate nearby indoor and outdoor subjects when the natural light is subdued. Therefore, relatively short exposures are used most of the time. This is by design, since a hand-held photo could easily become blurred by camera or subject motion if the exposure time was too long.

Film manufacturers obviously seek to optimize the performance of their product for the short exposure times which are most commonly used. This means that film performance may be less than optimum when a long exposure time is involved. Where the exposure time approaches an hour, some films may have only a fraction of the sensitivity their ISO number would indicate. In addition, reciprocity failure in some color films can result in a color shift due to nonuniform changes in spectral sensitivity versus exposure time. With some film, slight reciprocity failure can be detected in exposures as short as $\frac{1}{4}$ second.

It is relatively easy to quantify reciprocity failure and take some steps to counteract it. Most often, the best way to compensate is to just use a faster f/ratio or a longer exposure time. You can test a film for reciprocity failure, but there is usually little point in performing your own test because new films are constantly being introduced. In addition, you can easily learn about a given film by reading articles in printed publications or on the Internet. In the meantime, you can just start shooting your own astrophotos and recording your exposure data. The appendix of this book includes an astrophotography guide that should get you off to a good start. The longer exposure recommendations compensate for a modest degree of reciprocity failure. I developed the exposure guide by trial and error and used up a lot of film in the process. You may save a lot of film if you use it.

If you still want to test a film yourself and lack a suitable artificial light source, a good way to do it is to take a series of guided or unguided piggyback exposures of the night sky from a dark location and use combinations of f/ratios and exposure times which will provide the same exposure value. For instance, if you know that a film works well for piggyback photography when exposed for 8 minutes at f/2, then you can evaluate its reciprocity failure by taking additional piggyback exposures of 15 minutes (16 minutes if you want to be exact) at f/2.8, 30 minutes at f/4, 1 hour at f/5.6, etc.

In addition, you can take exposures of 90 minutes at f/5.6 and 2 hours at f/5/6. If the shot that was exposed for 1 hour at f/5.6 looks identical to the previously mentioned 8 minute at f/2 shot, then the reciprocity failure between the 8 minute and 1 hour exposure range is negligible. If the 1 hour shot is less dense (in the case of a negative) or more dense (in the case of a transparency) then the film has reciprocity failure. You can roughly quantify how much reciprocity failure the film has by comparing the 8 minute shot to the shots taken at 90 minutes and 2 hours at f/5.6. It is important to note that some films may have such severe reciprocity failure that it could create a "runaway" situation in which it is impossible to adequately expose a dim deep-sky object through a telescope having a slow f/ratio.

1.5.2.1 Compensating for Reciprocity Failure

The effects of reciprocity failure can be minimized in a number of ways: by selecting a film which has relatively little reciprocity failure; by "hypering" the film in appropriate gasses; or by using a cold camera which keeps the film at a low temperature during the exposure. Each approach has its own unique advantages and disadvantages.

A cold camera can increase the efficiency of typical films during long exposures. Some use thermoelectric (TE) coolers, but the most affordable ones use dry ice to cool the film. Most units have a metal plate of high thermal conductivity such as aluminum or copper in contact with the film to efficiently conduct heat from it. Cooling the film makes its surface susceptible to condensation and ice, but well designed cameras can prevent this from happening on the film's emulsion. Simple cold cameras have a thick transparent plug which is in contact with the film's front surface. More complex units either have a vacuum chamber in front of the film or a sealed chamber filled with a dry gas such as dry nitrogen.

Cold cameras are not particularly easy to use. Most can only be set up for one shot for every time they are cooled down. After each picture, you have to allow the film to warm up before you advance it (or change it if you are using film chips) in order to keep it from cracking and to prevent water ice from forming on the emulsion. In spite of these problems, there are benefits to using a cold camera. In my own experience, a cold camera has provided good results without discoloring the film as much as some gas hypering techniques do.

Gas film hypering is a very popular means of reducing reciprocity failure. Hypering is a process which must be completed before you take your picture, but it provides the advantage of allowing you to shoot your astrophotos with a conventional camera body and with no delay between exposures. Commercial film hypering equipment is available from various sources (most notably Lumicon), as is hypered film. Home grown film hypering tanks are another option if you know how to design and build one. With most film, gas hypering can shorten the required exposure time for dim objects by a factor or two or even substantially more, with the results typically being most dramatic for high resolution black and white films.

Hypering can be accomplished with various gasses and at a variety of pressures, times, and temperatures, and people are constantly developing new techniques. Some people use hydrogen and nothing else, while others wisely combine it with a dry, inert gas for obvious safety reasons. One thing that many agree on is that it is beneficial to pull a vacuum in the hypering chamber before introducing the forming gas. Where the hypering time is relatively short, it can be useful to spool the film on to a developing reel while it is in the chamber. If the film will be hypered over a couple of days or so, it does not seem to make as much difference whether the film is spooled onto a reel or left in the canister.

Drawbacks to gas film hypering include possible discoloration or apparent fogging of the film, typically of the part having the least exposure. This can be more likely to occur if you use a high temperature (more than about 38 degrees Celsius) in your hypering tank. Other disadvantages of hypering include a loss in sensitivity over time, but this can be minimized by limiting the amount of time the film is exposed to open air prior to taking your exposures. It is usually best to shoot your pictures within a day or two of the time you finish hypering your film.

To preserve the effects of hypering, some people leave their film in a hypering tank until they reach their observing site. Others go so far as to keep their film in a dry environment even while shooting an astrophoto by pumping dry gas into their camera. Who knows; maybe someone will go a step further and come up with a hypering tank that's also a camera!

Other techniques have been tried for low light photography, but not all are useful for astrophotography. One technique is controlled fogging, where the film is fogged (or "pre-exposed") to a level at or just below the threshold at which the fogging will just barely be visible on the processed film. This can allow dim subjects to more easily register on film, given a certain exposure value; however, the background sky glow at most observing sites already tends to brighten a deep sky photo more than many would like.

1.5.3 Film Formats

Films are available in different sizes, each for use in cameras having various image formats. Popular films and cameras are available for small formats such as the 17 mm wide format of APS cameras and the 24 × 36 mm format of popular 35 mm cameras; medium formats such as 6 × 4.5 cm, 6 × 6 cm, 6 × 7 cm, and 6 × 9 cm for 120 or 220 roll film or 70 mm film; and 4 × 5 inch and larger formats which commonly utilize sheet films.

The actual format size is typically smaller than the film itself, and it is usually a good idea not to fill the entire format up with your subject because most photo lab equipment does not quite image the entire format area on your prints. This slight image cropping is usually permitted in order to compensate for slight misalignment of film which can occur in an enlarger or slide mount. For example:

* A 24 × 36 mm format is common for full frame 35 mm cameras, but only about the central 22.5 × 33 mm of the image will be shown on the 9 × 13 cm prints made by most photo labs.
* The actual image area of the 6 × 45 cm format is about 39 × 56 mm, but typical prints may only include the central 36 × 52 mm.
* The actual image area of the 6 × 7 cm format is 56 × 70 mm, but typical 10 × 12.5 cm prints may only include the central 52 × 65 mm.
* The utilized area of the 4 × 5 inch format is only about 90 × 112 mm.

35 mm cameras are used by many amateur astrophotographers. The 35 mm figure indicates the approximate width of the film and has nothing to do with dimensions of the image format. Most 35 mm cameras provide a 24 × 36 mm image on the film, but a few (very few) specialized "half frame" cameras produce images of about 18 × 24 mm.

A larger image format usually provides a better picture due to the fact that a large original image does not have to be enlarged as much as a smaller original image. For instance: to get a 20 × 25 cm print, the

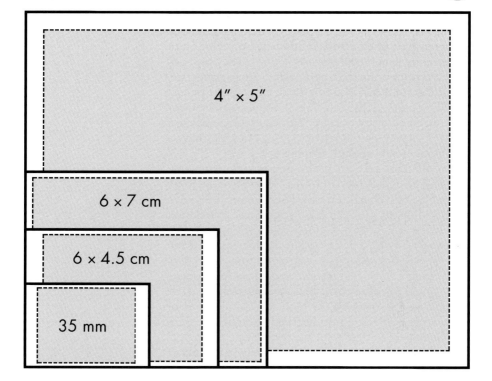

4" × 5"

6 × 7 cm

6 × 4.5 cm

35 mm

Figure 1.4. Comparison of various film formats. Dashed lines show the portion of the format typically utilized for commercial 20 × 25 cm enlargements.

utilized 52 × 65 mm area from a 6 × 7 format image on 120 film would only have to be enlarged 3.8×, while the utilized 22.5 × 28 mm part of an image on 35 mm film would have to be enlarged 8.9×. When a film image is enlarged, film grain, highlight blooming due to halation, and other artifacts are also enlarged to the same degree. This is part of why star images tend to look bloated in relatively extreme enlargements.

1.5.4 Film Grain

Photographic film is coated with an emulsion which contains all of the image information. This emulsion is relatively delicate, so all original negatives and transparencies should be handled with due care. Grain is a microscopic structure in the processed emulsion of film images. In some films, a magnified view of the grain structure looks similar to the randomized output of a color ink jet printer when a low print quality is selected.

Some fast films have a larger grain structure than slow films because their higher effective speed is achieved partly by virtue of allowing the grain over relatively large area of the film surface to clump together. This permits a given amount of exposure to influence the film more significantly than it could have otherwise. The drawback for this type of brute force approach to film sensitivity is that the grain tends to clump in all parts of the image, including parts of the image which should be of a virtually constant density or color. This can give the image a "grainy" appearance. Relatively grainy films may have a resolution of only about 1/40 mm or so.

Not all films rely on grain clumping to the same degree. Those that rely on it least often have the most favorable grain characteristics. Some newer films have small and tightly grouped grain structures. These grain structures usually have a more consistent appearance than the relatively clumpy grain in older films of the same speed, resulting in better resolution and a less grainy image. Such characteristics can lead to an image having finer or more consistent looking grain, a lower contrast grain structure, or both. For some types of images, the relative contrast between the grain structure and the subject matter in the image can be just as important as the actual size of the grain. Many new films of moderate speed have a resolution of 1/60 mm or better, and slower films can usually resolve even finer detail.

1.5.5 Halation

Halation is an undesirable characteristic that causes stars and other bright light sources to appear bloated in film images. It is caused by a number of factors, including light scattering within the emulsion; and from light passing through the emulsion, reflecting off of the back side of the film, and re-exposing the emulsion from the back side. In the case of a very bright star image, repeated reflections within the film base can easily cause its image to exceed a diameter of half a millimeter on the film.

For a given film, exposure value, point light source, and aperture, halation typically produces a defect of about the same absolute size, regardless of the film format used. In many cases, image blooming due to halation tends to be nonlinear with exposure value, so blooming can increase somewhat marginally as the exposure value is increased. This is yet another reason why a larger format image (with a correspondingly larger image scale) tends to look better; foreground star images appear smaller in relation to the subject.

1.6 Image Quality

Image quality is often defined in terms of sharpness, but there are other factors which are equally important. Some of these are contrast and color saturation. Even if an image is sharp, low contrast can make it look washed out.

1.6.1 Contrast

Low contrast can be caused by lens aberrations, optical design, excessive dirt or scratches on optical surfaces, too large a secondary obstruction (such as that in a fast Cassegrain telescope), flare, unwanted reflections, bad or no antireflection coatings, bad reflective coatings, atmospheric conditions, attributes of your subject, or other factors.

Flare is a common problem which occurs mostly on the surface of transparent material. You can see the undesirable effect flare has on contrast when you look through your windshield as you drive your car toward the rising or setting sun. You can also see that the flare tends to be reduced when your windshield is clean. Most good camera lenses have multilayer antireflection coatings which minimize flare and increase contrast.

Clean optical surfaces are important for photographic lenses. Image contrast is typically highest when the optics are clean; however, one should not be overzealous in cleaning optics, because it is easy to scratch a lens while cleaning it. A few scratches of various sizes can be tolerated, but a large number of microscopic scratches can be just as detrimental to your image as a lot of dirt on the lens – and scratches can't be removed!

A lot of image degradation from dirt and scratches is the result of diffraction. Since diffraction occurs at the boundary of an obstruction, the amount of area covered by dirt or scratches can be less important than the combined length of their boundaries. For instance, the boundary for a seemingly large chip or dust particle may add up to a few millimeters at most. This can be tolerated on most normal and telephoto lenses; particularly when you consider that some mirror lenses and Cassegrain telescopes have a central obstruction of more than $\frac{1}{3}$ of their diameter, yet still work fine. By contrast, the boundaries of hundreds of small scratches can add up to several meters! Some "soft focus" attachments are actually made up of a myriad of small etched lines. You don't want to turn your lens into one of these by overzealously cleaning it.

Regardless of what influences the contrast of a photo, contrast has a lot to do with how fine a detail

Figure 1.5. Upper left: A few large scratches on the front of this 300 mm ED Nikkor telephoto lens have virtually no effect on photos taken through it. Lower left: The many fine scratches (by arrow) on the rear element of this well-used 105 mm f/2.5 Nikkor lens affect the image more than all of the other flaws combined. These scratches alone cause enough diffraction to significantly affect some astronomical images. Right: The crescent moon in front of the Hyades star cluster, taken through the 105 mm lens shown at left. Spikes on Venus (at upper right) are caused by diffraction from aperture blades in the lens, which was stopped down to f/4 for this 20 second exposure on Ektachrone 200 slide film. The long streak through the moon is caused by diffraction from many small scratches on the rear element of the lens. The scratches are nearly parallel to each other, so a significant amount of light was diffracted into a streak perpendicular to them. If the scratches had instead been oriented in many different directions, the effect would have been general blooming around the lunar image. This lens works fine for pictures of normal daytime subjects, showing that astrophotography can have more demanding optical requirements. The consolation is that I got the lens for only $40 due to its optical and mechanical condition. I was able to repair its mechanical problems, but scratched optical surfaces are not a "fix it yourself" item.

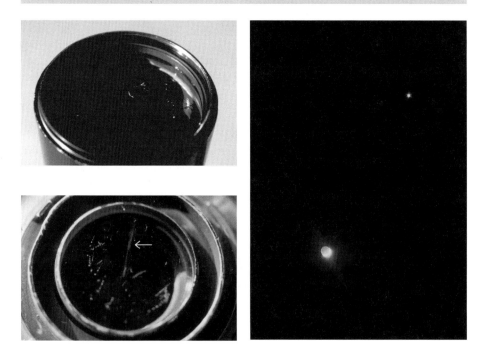

can be effectively imaged. A low contrast subject is typically harder to image sharply than a high contrast one, so in general the high contrast subject will tend to facilitate a finer apparent resolution. For instance, a high contrast boundary such as the limb of the moon will tend to appear to be imaged more sharply than low contrast objects such as nebulae or the belts on Jupiter. Accordingly, it is a good idea to get an image of a sufficiently bright but low contrast subject (such as the surface of a planet) that is substantially larger than an image in which the sharpness would just barely be limited by film characteristics.

Contrast can also be affected by optical design. Low contrast lenses usually do not produce as good an image as a high contrast optics of the same sharpness. The combined effects of sharpness (or resolution) and contrast are addressed in the "modulation transfer function" (MTF) of an optical system; however, MTF cannot be quantified for a given lens without specialized equipment and techniques. Therefore, this book will deal with contrast and resolution separately, with resolution being expressed in terms of line pairs (one light line and one dark line) per millimeter. Optics are not all the same. In the real world, many lenses and telescopes do not attain anywhere near theoretically perfect performance, as will be addressed in later sections.

1.6.2 Color Saturation

Color saturation is closely related to contrast, being influenced by many of the same factors. The difference is that low color saturation is more difficult to correct than low contrast. The contrast of a black and white image can be increased by printing it on high contrast paper. Contrast in a color image can be corrected with sandwiching and masking, but this usually does little to improve color saturation unless one gets into the complexity of making color masks. Digital image processing can obviously increase color saturation, but excessively increasing the saturation of a low saturation original can make the end result look artificial.

The best way get good color saturation in a final image is to have an original with good color saturation. This usually means that you need to use good optics and be relatively conservative in regard to film speed. Fast film will shorten your exposure time, but too fast a film can provide a mediocre image. Even in cases where a fast film maintains high resolution, there is usually some penalty in terms of dynamic range and color saturation. One of the most obvious losses in comparatively fast film is in the subtle color of stars.

Today's ISO 400 and faster color films are a lot better than they used to be, but the fastest films still lose some color information. ISO 800 to 1600 films have even less color saturation and resolution, though they still provide acceptable results for some subjects. Films in the ISO 3200 range can be fine for photos taken through a telescope, but slower film can provide more rewarding piggyback images. For example, even my oldest Ektachrome 200 piggyback shots reveal a lot of subtle color in various stars, and today's version of the same film is even better.

Going to medium format and using a longer focal length lens can improve the angular resolution you get with a given film, but it won't compensate for the lower color saturation of a fast film. If you use a slow to moderate speed film with a good camera lens, you are more likely to get better results in regard to color saturation.

1.6.3 Sharpness

It may always seem desirable to have the best possible image quality in terms of sharpness, but small prints of some subjects taken with "good" optics may be almost indistinguishable from those taken with "excellent" optics. In evaluating the overall image quality you need for various applications (and how much you spend to achieve it), it is important to consider the size of prints you will want, the distance from which prints (or projected images in the case of slides) will be observed, and how much money you can throw at the task.

A good rule of thumb for very high quality images is to shoot for a resolution of about $\frac{1}{4}$ mm on moderate size wall-mounted prints. If the prints will be observed from close up (as may be the case if they are in a photo album) a resolution of about 1/10 mm will look really sharp, but less resolution is acceptable in most cases. These are only suggested values to shoot for if you want really sharp results. Most people won't see the difference between such prints and those of higher resolution. In some cases, prints of substantially lower resolution will still look good. For reference, the half-tones in this book are screened at 6 dots per millimeter or 150 dots per inch. This resolution allows the pictures to look good, though they are still somewhat inferior to the resolution of a good photographic print.

If you are using a 35 mm camera and only need to get really good 9 × 13 cm prints, your film image will only have to be enlarged four times. This means that the image on your film must be four times sharper than the required sharpness of your print, so a resolution of 1/40 mm would be required of the film image if you want a resolution of 1/10 mm on your print. This is not too hard to achieve with good optics. A telescope used at prime (i.e. direct) focus will usually exceed this resolution at the center of the image; and amazing as it may sound, it may do this even if its optics have substantial imperfections!

Even though slide projection screens are larger than most prints, an original slide having a resolution of 1/40 mm is usually adequate for projection because few projectors have optics capable of resolving much more detail than that. I have seen some projector optics that were so bad it made me wonder if the bottom of a soda bottle wasn't being used for the lens. It obviously won't hurt if the slide is sharper than the projection lens because the effects of slide sharpness and projection lens resolution are cumulative.

If you want to end up with a 20 × 25 cm print which meets the same sharp 1/10 mm resolution requirement because it will be observed up close, the original 35 mm format image will have to be about twice as good as that required for a 9 × 13 cm print. It can be challenging to get a 35 mm image that is sharp enough to hold up this well under so much enlargement; however, a good 20 × 25 cm print would not be a problem for a good medium format image, since the medium format image would not have to be enlarged as much.

Given the same film, exposure value, and field of view, a medium format camera can provide an image having about twice the detail of a 35 mm photo; however, medium format lenses usually have slower f/ratio than a 35 mm format lens with the same field of view. This requires a longer exposure, but many photographers believe that the improved results (particularly for wide field photos) are well worth the extra effort. The down side to medium format is that the equipment can be relatively expensive, and there are many subjects (such as those having too small an angular size to fill the format with available or affordable optics) for which a medium format camera would provide no benefit over a smaller format one.

A larger format is only useful when the subject can occupy most of the image area. Therefore, a 35 mm format camera is just as effective as medium format one when using a typical amateur telescope to photograph astronomical objects having an angular size substantially smaller than the moon.

It is important to point out that no matter how sharp your optics are, some features will be imaged more sharply than others. Halation and other effects will invariably cause bright star images to be far larger (perhaps many times larger) than the resolution criteria would indicate. In the case of some objects, less than perfect image quality can and must be tolerated because of limitations imposed by vibration, optical performance, atmospheric "seeing" conditions, and exposure time. These factors will be addressed further in later sections.

In the end, it is not the camera that matters most, but the person behind it. If you utilize proper photographic techniques, you can achieve relatively impressive results in some types of astrophotography even if you use modest equipment.

Chapter 2

Astrophotography Basics

Astrophotography presents many unique challenges and rewards. A few astronomical objects and events are observable during the day or can be photographed with wide angle lenses, but most have a small angular size and are so dim that they are visible only at night. When photographing such dim and distant subjects, good high resolution optics, fast film, a long focal length, and a large aperture are advantageous. This usually implies the use of a large telescope, but there are ways to get relatively good images with a small telescope. Selection of a good observation site is also important, because city lights, clouds, or fog can wash out or obscure one's view of the night sky.

2.1 Types of Telescopes

Telescopes are available in a wide variety of designs and sizes. Those used by amateur astronomers can range from pocket size to larger than a refrigerator and cost anywhere from the equivalent of a few dollars (second hand) to the price of a house.

Which telescope is best for you will depend on what you want to use it for, so it is a good idea to write down what you want a telescope to do (and cost) before you get one. Your requirements should drive what telescope(s) and accessories you get. This is better than the opposite approach where you get the first telescope you

come across, then discover that it won't work very well for your purposes. Your requirements are more important to YOU than the type of telescope your friends are using or what looks good in a flashy brochure or magazine advertisement.

Your requirements may have a great deal of influence on how much you need to spend for a telescope, but you can always look for a deal on a used telescope if you want to get the most for your money. If you want to emphasize planetary photography, good optical performance will be particularly important. If instead you want to photograph a variety of very dim objects having a small angular size (such as most galaxies), you will need a relatively large telescope, but its optics won't necessarily have to be as good as what is required for good planetary imaging; however, you will need a very good mounting, motorized sidereal drive, and the proper guiding accessories. If you only want to photograph certain "showpiece" objects such as the great nebula in Orion (M42) or the Lagoon nebula (M8), you will not need a particularly large or expensive telescope, though your exposure time for a given image size will be shorter if the telescope has a larger aperture.

If you only want to photograph wide areas of the night sky, you may not need any telescope at all; other than possibly a small guide scope to verify tracking of your camera. For wide field photography through camera lenses, a relatively simple home made tracking mount (often called a "barn door" mount) may be adequate. If you only want to photograph phases of the moon, virtually any good telescope (or sufficiently long focal length camera lens) will do.

It is important not to always go by a telescope manufacturer's claims for magnification and the like. If a telescope is made well, the diameter (i.e. aperture) of its optics is what will determine the maximum practical visual magnification. An often used rule of thumb is

Figure 2.1.
Left: Refracting telescope on German equatorial mount. Center: Newtonian telescope on alt-azimuth mount. Right: Cassegrain telescope on fork mount. Appropriate tilting of a fork mount allows its polar axis to be parallel with the earth's axis of rotation. On the back of the Cassegrain telescope is my home made prototype of the "VersAgonal", a multiple function flip mirror I invented back in 1982. A subsequent commercial version is shown on the back of the refractor telescope.

that a magnification of about 20× per centimeter of aperture is the maximum that will look sharp in a good telescope. This can vary according to your requirements. For my own tastes, this value may range between about 15 and 27× per centimeter, depending on the subject, atmospheric seeing conditions, and the relative size of a telescope's secondary obstruction, if any.

Some telescopes as small as 60 mm aperture are purported to provide a magnification as high as 450×. Here, the manufacturer is just blowing smoke. Such a telescope may include a minuscule eyepiece which will provide 450× magnification when used in conjunction with a Barlow lens, but such an image won't look sharp. If you try to look at a planet at 450× through such a telescope, all you'll see is a fuzz ball. In addition, many of these small mass-market telescopes have spindly mountings which are frustrating to use. Such a telescope and mount may provide the resolution and stability for a decent image at 100–150× (if there is no breeze to shake its mount), but that's about it.

A more important consideration is that you typically do not need a lot of magnification unless you are observing or photographing planets or double stars. Indeed, the problem with some telescopes is that they do not provide a low enough visual magnification for wide field subjects. This problem can usually be corrected by getting a lower power (longer focal length) eyepiece.

The objective lens of a common small telescope may have s focal length between 400 mm and 1000 mm. This is between 8 and 20 times longer than the normal lens of a 35 mm camera. You can usually take pictures of the moon through a small telescope if you have a proper adapter or use other methods which are covered later in this book

Following are brief descriptions of a few basic types of telescopes.

2.1.1 The Refractor

A refractor telescope uses only lenses to form the image. In a simple refractor, all of the objective elements are at the front and an eyepiece is at the back, giving it the familiar appearance that comes to mind when most think of a telescope. For visual observation, a right-angle reflector attachment (star diagonal) is often used at the back. This provides easier access to an eyepiece when the telescope is pointed upward. For astrophotography, some means of seeing the camera finder from a right-angle position may be required if you are to comfortably take pictures.

More than one lens must be used for the objective of a good refracting telescope because a single lens causes light of different wavelengths (colors) to be in focus at different distances from the lens. The index of refraction for a given optical glass is effectively higher for a shorter wavelength, so a short wavelength such as blue is brought to focus closer to a simple lens than a longer wavelength such as red. Most early refractors (and many current ones) are achromatic designs having two elements.

The proper use of two optical elements having different refractive indices allows the optical system to bring two colors of the spectrum (usually red and blue) to a common focus. They do not bring all colors to a common focus, but they are a dramatic improvement over a single lens. The focus for red and blue light is at a different distance from an achromatic objective than that for greenish yellow light (the color to which the eye is most sensitive), but the use of special glass can lessen the focus difference.

Refractors optimized for visual use have historically been designed to bring the above mentioned blue and red light to a common focus, at the expense of allowing a visible blur of out of focus violet color around bright objects. This violet blur is not excessively distracting for visual observation with most small to moderate size refracting telescopes; however, most photographic film has a greater relative sensitivity to violet (and even ultraviolet) light. Therefore, most film will record the out of focus violet color to a significant degree, resulting in color photos having an obvious violet glow around each bright star image.

Apochromatic telescopes bring three colors of light to a common focus. These typically have three or more elements, but it is important to point out that not all three-element telescopes are true apochromats; some only bring two colors to a common focus. Either way, a well designed three-element telescope will tend to bring all visible colors closer to the same focus than would a simple achromat, resulting in a sharper photographic image. The proper use of special optical materials (particularly fluorite and ED glass) can provide even better images.

Some three-element refractors are optimized for astrophotography. Instead of bringing red and blue to a common focus, these telescopes (particularly those which are not true apochromats) are corrected to bring red and a color between blue and violet to a common focus. This can significantly reduce the size of the violet blur in film images, but it does not always improve visual performance by the same degree. Therefore, those wanting to emphasize visual observation of planets may want to exercise moderation in how fast a refractor they select, even if it has three elements.

Refractor telescopes offer the advantage of an aperture that is free of obstructions. This in turn can result in sharper and brighter images for a given aperture. Refractors with a large diameter focuser have less vignetting than most other types of telescopes, and some can even be used effectively with medium format cameras.

Compact small to moderate aperture apochromatic refractors have relatively short focal lengths, making them ideal multiple purpose instruments for both wide and narrow field photography of astronomical objects as well as for observation and photography of wildlife, air shows, auto races, and many other subjects. All of this versatility comes at a price because refractors (particularly apochromats) are typically the most expensive type of telescope per centimeter of aperture.

Some high end ED telephoto camera lenses are good enough to be used as telescopes. They tend to be very compact for their aperture and most have a smooth focus control. They also include an adjustable aperture, which can be very useful in photographing terrestrial subjects.

2.1.2 The Newtonian Reflector

The Newtonian telescope is an elegant optical system consisting of a concave paraboloidal primary mirror and a smaller flat diagonal mirror. The focuser is usually on the side of the tube, near the front. The diagonal mirror (sometimes called a secondary mirror) is typically located directly in front of the primary mirror, but there are some off-axis designs based on the Newtonian principle in which the diagonal mirror does not obstruct the aperture. Other related designs may have correcting optics at the front or a correcting lens or Barlow lens in the focuser tube.

A Newtonian is typically the least expensive telescope per centimeter of aperture. It offers many advantages, including an integral right-angle focuser position which eliminates the need for a star diagonal or a right-angle attachment for one's camera finder when the telescope is pointed near the zenith. Since the eyepiece and camera are near the front end of the tube, the mounting structure can be lower, which in turn can increase stability.

The Newtonian telescope can be made with a fast f/ratio primary mirror, but such a mirror is more subject to coma, an aberration which produces teardrop or "v" shaped off-axis star images. Fortunately, coma corrector lenses work very well for astrophotography. Most will provide good off-axis images even when used with a Newtonian having an f/ratio as fast as f/4. Slower Newtonian telescopes have less of a problem with coma, and one working at f/6 or slower can provide images having very little visible coma on the 35 mm format, even without a coma corrector.

It's amazing what can be done with a Newtonian, particularly if one is conservative in regard to f/ratio. For wide field imaging, a Newtonian telescope and coma corrector can be on a level playing field with an expensive apochromatic refractor. Here, it may be enlightening to point out that an f/6 Newtonian is often considered "slow", while a refractor of the same f/ratio would be considered "fast". In addition, an f/6 Newtonian with enhanced coatings and a small diagonal is almost as efficient as an f/6 refractor of the same aperture; and the Newtonian usually costs a lot less!

Large film formats can be utilized with a Newtonian if the diagonal mirror and focuser are large enough, if the f/ratio is sufficiently slow or appropriate corrector optics are used, and/or if the primary mirror is optimized to reduce coma at the expense of some reduction central performance. These configurations are usually acceptable for wide field astrophotography.

2.1.3 The Cassegrain Reflector

A Cassegrain telescope consists of a concave primary mirror and a smaller secondary mirror which is usually convex. The secondary mirror is typically located directly in front of the primary mirror and the focal surface is behind a hole in the center of the primary mirror. As with the refractor, right-angle viewing attachments may be useful for visual and photographic applications when the telescope is pointed upward.

Cassegrain telescopes are made in various designs which have different figures on their primary and secondary mirrors. Some popular Cassegrain telescopes also have one or more built-in corrector lenses. In this case, the telescope description usually includes the name of the corrector design. For instance, a Cassegrain telescope having a Schmidt corrector is called a "Schmidt–Cassegrain" telescope, or SCT. There are also some off-axis Cassegrain designs in which the secondary mirror does not obstruct the aperture.

A Cassegrain telescope's secondary reflector typically acts to narrow the convergence angle of the cone of light from the primary mirror, providing an effective focal length which is radically longer than that of the primary mirror alone. For example, the primary mirror may have an f/ratio of f/2, but the curvature of the secondary mirror may slow the working f/ratio at Cassegrain focus down to f/10 or so. Combined with the folded light path, this allows a Cassegrain telescope to be very compact for its focal length.

Disadvantages of the Cassegrain include a higher cost per centimeter of aperture (as compared to a typical Newtonian) and a secondary obstruction which is larger than that of a Newtonian having the same f/ratio. Too large an obstruction can affect the quality of high magnification images, but it is not usually a problem for wide field applications. A Cassegrain telescope having an f/ratio of 8 or slower is typically better for general use than a faster one due to its smaller relative obstruction size. A significantly faster f/ratio system would only be advantageous if deep sky observation and photography is emphasized over planetary imaging.

2.1.4 The Schmidt Camera

The optics of a Schmidt camera consist of a concave primary mirror and a front Schmidt corrector lens which may be somewhat smaller. This difference in size between the entrance aperture and primary mirror facilitates a relatively even image brightness over the entire film format. A special film holder having provision to curve the film to match the focal surface is typically mounted directly in front of the mirror. Schmidt cameras having an inside focal surface such as this are not used for visual observation. Therefore, some do not consider them "telescopes" in the strictest sense of the word.

A few popular Schmidt cameras have film holders which accept only small chips of film, but some of these can be modified to accept a roll of film. A simple form of such a modification involves using a second film cartridge as a take up spool, then supporting both film cartridges on a wire frame. The film is manually advanced between exposures by spinning the take up spool, usually in a light-proof changing bag. Since the film remains uncut, it can be processed by conventional photo labs. This type of roll film adapter is also useful for some cold cameras.

Schmidt cameras provide many advantages for astrophotography. First, they provide a wide field image (up to several degrees) of exceptional quality. They also tend to have a very fast f/ratio, with f/1.5 not being uncommon. Such a fast f/ratio makes the Schmidt camera very useful for photographing nebulae and other dim objects, so long as the subject has a sufficiently large angular size. They are also useful for photographing objects such as comets which may have a rapid proper motion in front of background stars. When a comet has a fast proper motion, one must choose between tracking on it or the background stars. The short exposure facilitated by the Schmidt camera minimizes streaking of objects in the photo which are not properly tracked.

The fast f/ratio of a Schmidt camera makes it relatively expensive to manufacture. Part of the cost is due to the fact that a low expansion material called Invar (TM) has historically been used to provide acceptably consistent positioning of the film holder when the camera is used in environments of various ambient temperatures. Schmidt cameras have some very good attributes, but they are not used by most amateur astronomers due to their cost and relatively specialized application.

Some slower Schmidt camera designs have a diagonal mirror, additional correcting optics, and a side focal surface position which can allow the use of a conventional camera body or even visual observation with an eyepiece. Some of these use a true Schmidt corrector, while others more accurately fit the description of a "souped up" Newtonian.

2.1.5 Unusual Telescopes

Telescopes are made in a variety of other designs. Some, such as the Tri-Schiefspiegler, are reflecting designs having an unobstructed aperture. Others are convertible from a Newtonian to a Cassegrain. Regardless of the traditional pros and cons relating to various optical designs, the bottom line is how well a given system will work for your own applications.

2.2 Telescope Mountings and Drives

This section provides a brief overview of why a sidereal drive is required for some types of astrophotography, as well as covering various types of telescope mountings. We will not actually get into the physics of celestial mechanics, but the impact celestial mechanics can have on astrophotography will be addressed.

The earth rotates in relation to the stars, so to the terrestrial observer celestial objects appear to move in relation to the horizon. Obviously, if the exposure time is long enough, celestial objects will appear to be streaked in a picture unless the camera is moved in a way that will allow it to track them. The apparent motion of celestial objects at any latitude other than the north or south pole will vary according to azimuth, and the camera's tracking motion must match the motion of the subject.

2.2.1 The Equatorial Mount

An equatorial mount is the most common means for tracking celestial objects. When combined with a sidereal drive, an equatorial mount will effectively cancel out the earth's rotation by rotating at the same rate as the earth but in the opposite direction. Parallax related to the distance between the observer and the earth's axis does not pose a problem when tracking the stars because they are at such a great distance.

There are ways to track the stars that don't involve than using a true equatorial mount. Computer controlled altitude azimuth telescope mounts have become increasingly popular and effective. Using an alt-azimuth mount is both familiar and intuitive, since one axis moves up and down and the other moves from side to side, just like the motions on a typical camera tripod.

A computer controlled alt-azimuth telescope mount
can provide a sidereal drive rate without the need for a
true equatorial mount by slowly tracking in both axes.
Some of these units utilize a fork mount. When used
vertically in an alt-azimuth position, a fork mount can
provide better access to the eyepiece than it would if
used in equatorial mode, where it is tilted so its
azimuth axis is parallel with the earth's axis of rotation.
When a fork mount is tilted in this way, the azimuth
axis becomes the polar axis.

Alt-azimuth telescopes which track by moving both
axes require a camera rotator (sometimes called a field
de-rotator) for deep sky astrophotography in order to
prevent visible field rotation during long exposures.
These rotator units are currently available for use
between the telescope and the camera; however, rota-
tors for piggyback mounted cameras are not widely
available, at least not yet. In the meantime, it may be
possible to adapt a custom piggyback camera bracket
to existing rotator units.

For simplicity, most examples in this section will
deal with a true equatorial mount which has sidereal
drive. Such a mount has two axes; a polar axis which is
positioned to be parallel with the earth's polar axis, and
a declination axis which is at a right-angle to the polar
axis. To track celestial objects, the polar axis rotates at
the same rate as the earth's rotation, but in the opposite direction. This effectively cancels out the rotation
of the earth with the simple motion of a single axis,
eliminating the need for a camera rotator.

A variety of commercial equatorial mounts are available, including the German equatorial mount, the fork
mount, and the extended polar axis mount. A given
German equatorial mount can be designed for compat-
ibility with a wide variety of telescopes. The telescope
tube is offset to one side of the declination bearings,
necessitating the use of a counterweight to balance the
mount in right ascension. This increases the overall
weight of the mount, but the offset telescope position
has the advantage of providing good access to the eyepiece regardless of where the telescope is pointed.

The fork mount is often utilized for Cassegrain telescopes because the telescope can be centered between
the declination bearings, eliminating the need for a
heavy counterweight. This typically makes the fork
mount lighter for a given telescope, but access to the
eyepiece can be difficult with some commercial units
(particularly those having an offset wedge) when they

are pointed toward an area between the zenith and the celestial pole. Commercial fork mounts also tend not to be very adaptable to different telescopes, and some can even limit declination coverage, a potential problem when one lives relatively near the equator.

Many astrophotographers have resorted to building their own equatorial mounts and tracking platforms. These can be made in a variety of shapes, sizes, and configurations, and they need not be made entirely from scratch; existing bearing assemblies from various other gadgets can be used for at least the polar axis. Some home made mounts are low profile platforms which can be used under simple alt-azimuth mounts like those used on many inexpensive Newtonian (or "Dobsonian") telescopes. These platforms only provide continuous tracking for a limited time, but this is not usually a problem because tracking is only required while you are setting up for and taking your photos.

Positions of celestial objects are specified according to an imaginary spherical polar grid system. In principle, this celestial grid can be envisioned as corresponding to longitude and latitude on the earth, with the common point of reference being in the center of the sphere. Divisions around the polar axis of a celestial grid correspond to longitude in principle, but the celestial grid is calibrated in units of hours, minutes, and seconds of right ascension. Declination corresponds to latitude, or the distance toward or away from the equator. Declination is described in terms of degrees north or south of the celestial equator. It is interesting to note that if you set the declination to an angle equal to your latitude, the telescope will point straight up when it is pointed toward the meridian. The meridian is an imaginary line which extends from north to south in the sky and intersects the zenith, or the point directly overhead.

2.2.2 Compensating for Subject Motion

One thing that celestial objects (other than those at the celestial pole) have in common is that they appear to move across the sky as the earth rotates. The motion is most pronounced near the celestial equator because it is farthest from the poles. A star closer to the celestial pole will appear to move slower, with one at about

60 degrees declination moving half the rate of one at the celestial equator.

This apparent motion of celestial objects will cause smearing or streaking in photos shot from a fixed tripod. Such smearing will be visible in your image if the exposure is very long. The point at which smearing becomes visible in a photo is when it produces streaks or smears which are larger than a good percentage of the maximum resolution of your optics and film. Most celestial objects are so dim that they require exposures long enough to cause visible smearing if they are not tracked.

The maximum allowable exposure time for a given amount of image smear is determined by focal length, with a longer focal length being less tolerant of long untracked exposures. Which objects can be photographed without tracking is dependent on focal length, the required exposure time, and the required resolution. Fast film will facilitate photographing dimmer objects from a fixed tripod for a couple of reasons. First, fast film is more sensitive. Second, fast film typically has a lower resolution, which in turn means that more smearing can be tolerated before it becomes obvious in the image.

The resolution to shoot for will depend on the nature of the subject and the size of your print, slide projection, or computer display. A streak about 1/40 mm long on an original 35 mm negative will only be about 1/10 mm on a 9 × 13 cm print. It is safe to say that this amount of smearing will not be obvious to most people. If you are not too picky, a smear of even twice that amount may be acceptable for such a small print.

The maximum allowable exposure time versus focal length and acceptable smearing can be calculated without much difficulty. In addition, the relationship between lens focal length and allowable image smear is linear, so you need only calculate the smearing for one of your lenses, then just multiply the result by the proportional difference in focal length for other lenses. You can even "wing it" by simply comparing your lens focal length and desired resolution to one of the examples below.

Let's say that you want to know how long an exposure you can get away with if you want to limit smearing or streaking to 1/40 mm when a 50 mm lens is used on a fixed tripod.

First, we calculate the ratio between the focal length of your lens and the acceptable image smear: 0.025 mm smear / 50 mm focal length = 1:2000. This means that the subject can appear to drift up to an angle corresponding to this ratio during the exposure.

A known angular ratio can be divided by the acceptable drift ratio. For instance, a one arc second field of view is 1/206,265 of the distance from the telescope or lens. If we divide the 206,265 ratio by the acceptable drift angle ratio of 1/2000, we get 102 arc seconds: 206,265/2000 = 102.

A star on the celestial equator will move across the sky at a rate of about 15 seconds of arc (1/240 degree) per second of time, so the star will drift 1 second of arc in 1/15 of a second (0.0667 seconds) of time. To get the acceptable exposure time, just multiply 0.0667 seconds by 102 (for the 102 arc seconds of drift angle in the above paragraph): 0.0667 seconds of time × 102 = 6.8 seconds maximum allowable exposure time. There can be some fudge factor in this value because the miniscule difference in smearing between a 6 and 8 second exposure will probably be indistinguishable.

Now let's say that you are using a 200 mm lens instead of the 50 mm. The 200 mm lens is 4 times longer, so the maximum allowable exposure time will only be $\frac{1}{4}$ as long, or 1.7 seconds. The linear relationship between lens focal length, allowable smearing, and maximum focal length is only strictly true at the center of the image, but the discrepancy between the center and edge of the image would only be significant for extremely wide angle rectilinear camera lenses.

If you want to take longer exposures and still get good results, it will be necessary to track your subject during the exposure. This is where a sidereal drive comes in.

2.2.3 Benefits of a Sidereal Drive

As mentioned earlier, the purpose of a sidereal drive is to effectively cancel out rotation of the earth by rotating a telescope or camera at a rate and direction that is exactly opposite that of the earth's rotation. The term "sidereal" is typically used because it is the rate corresponding to the sidereal day, which is the time it takes the earth to rotate a full revolution in relation to the stars. Due to the earth's orbit around the sun, each year has one more sidereal day than it has solar days. Accordingly, a sidereal "day" is about four minutes shorter than the solar day. If you want to track a star, you would use the sidereal drive rate, which is about one revolution of the right ascension (R.A.)axis every 23.934 hours. If you want to track the sun, the solar drive rate would be used, which is one revolution every 24 hours.

Even though the sidereal day is only about four minutes shorter than a solar day, the required difference in tracking rate can be enough to make solar rate a problem when you want to track stars during a long exposure through a long focal length telescope. Relative to the stars, the sun appears to move eastward at just under one degree per day, and your telescope will do the same if it is tracking at the solar rate. If you use the solar drive rate during a one hour exposure of the night sky, the stars would be streaked almost 1/24 degree, or about 2.5 arc minutes. If the telescope has a focal length of 2000 mm, this would translate to a 1.5 mm streak on the final image.

You can easily correct for this much drive error with a drive rate corrector, but if you want to guide to an accuracy of 1/40 mm, you will have to make one drive rate correction an average of once a minute just to compensate for the difference in tracking rate. You can certainly guide a picture in this way, but things will be more optimum if your average drive rate is closer to the mark. Getting the correct drive rate is simply a matter of having the correct gearing and motor speed.

Even when the drive rate is correct, some drive corrections are still necessary. Atmospheric refraction, inaccurate polar alignment, imperfections in the drive mechanism, and proper motion of the subject (such as in the case of a comet) can all contribute to poor tracking. When making drive corrections, you will require some means to verify how well your telescope is tracking. This is usually accomplished by using a guide scope, off-axis guider, beam splitter, or other accessory to view a guide star that is in or near the area being photographed.

A good sidereal drive will permit you to track celestial objects for astrophotography, but a *stable* telescope or camera mount is just as important; it is even essential for quality astrophotography with a long focal length telescope. Emphasis is often placed on the quality of the drive gear, but the overall stability of the telescope mount is far more important. Drive rate errors occur in discrete axes and can be controlled with a drive corrector. Flexure and vibration of an unstable telescope mount can occur as rapid motions in multiple axes, making effective correction almost impossible. Therefore, the best solution is to use a mount having minimal flexure and vibration.

2.3 Types of Astrophotography

There are many basic types of astrophotography; so many that it is difficult to put them all into categories. Some are easy and some are tedious, but all are rewarding. Some types of astrophotography only require a camera and a tripod. Some require an equatorial mount or sidereal drive to track celestial objects, and some require even more equipment for actively guiding during an exposure.

Many consider telescopic deep sky photography of nebulae and galaxies to be the pinnacle of astrophotographic endeavor, but there are many other types of astrophotography which provide their own unique results. You can photograph the entire sky with a fisheye lens or a wide angle reflector; star fields and conjunctions with a normal lens; the moon, an eclipse, and larger nebulae with a telephoto lens or telescope; and planets and nebulae of small angular size with a telescope and auxiliary optics.

2.3.1 All-Sky Photography

All-sky photography is where you capture the entire sky in a single image. If the exposure time is relatively short, a good all-sky image can be captured without a sidereal drive. All-sky optics can be used to photograph a variety of subjects, including the Milky Way, the zodiacal light, meteor showers, aurora, brighter satellites, and even the shadow of the moon during a total solar eclipse.

If only the sky is your subject, you can image it with a fisheye lens or curved reflector having a 180 degree field of view. It is surprising how simple optics for all sky photography can be; convex hubcaps have been used for decades. Better all-sky optics include metal coated condenser lenses and fisheye lenses.

If you want to cover more than just the sky and get the entire horizon with some foreground objects in one image (to essentially capture a 360 degree panorama in a single picture), you will need a lens or reflector having a field of view which significantly *exceeds* 180 degrees. Some hubcaps, condenser lenses, and fisheye lenses will cover slightly more than 180 degrees, but the most interesting images are obtained with optics having a field of view approaching 260 degrees or more. This much coverage typically calls for a less common type of lens or reflector which has a lot of curvature. Ideally, such a reflector will have prolate aspheric figure such as a paraboloid or hyperboloid in order to minimize radial compression of the image near the edge of the field of view.

It is typically easiest to position the camera directly over the reflector, but other configurations are possible. One of these is a "Cassegrain" design, where the camera is positioned behind a hole in the convex reflector and a secondary mirror is used at the front. If you only want to photograph part of the horizon and a large part of the sky, you can instead tilt the reflector and mount the camera beside it.

To eliminate any distracting background around the reflector, you can either add a hood around the camera lens which blocks out everything except the reflector or you can use a dark cloth or piece of cardboard behind the reflector. If the reflector is placed on the ground, it is a good idea to set it on something like a towel. Using a rough cloth such as this can reduce the likelihood of a breeze blowing dust onto the reflector during your exposure.

An obvious concern in using a reflector is that the reflection of the camera or secondary mirror will be in the wide angle image. If the camera or other obstruction is farther from the reflector, it will be smaller in your picture. Therefore, you will get a smaller camera

Figure 2.2. Left: You can get all-sky photos by using a simple convex reflector under your camera. Right: An all-sky exposure of only a few minutes with an all-sky reflector can capture constellations, the Milky Way; and in this case, two Leonid meteors! Photo courtesy of Pierre-y Schwaar.

Figure 2.3. Left: An axial strut can eliminate side obscurations in all sky images. Right: Only a small "camera nebula" appears in the center of the sky in this fireworks photo which was taken with an axial strut reflector on a cloudy fourth of July in Estes Park, Colorado.

reflection if you use a normal lens to photograph the reflector than you would if you used a wide angle lens.

An even more important concern is that some structure must be used to support the camera (or the secondary mirror in a Cassegrain design), and it is desirable to minimize the impact such a structure will have on the image. Many astrophotographers use tripods or three vane spiders to hold a camera above the reflector. A spider can have a thin cross-section as seen from the center of the reflector, resulting in thinner obstructions than would be caused by conventional tripod legs. This is illustrated by Pierre Schwaar's photo of meteors in the fall sky (Fig. 2.2). Instead of a regular tripod, Pierre used thin profile steel bars to support his camera above a hubcap reflector.

A single low profile vane would have less effect on the image, but few people seem to use this approach. Other designs include a transparent dome or cylinder around the reflector. This is more or less invisible, but it can introduce distortion or flare if its surface is not properly figured or if it lacks antireflection coatings.

In the 1970s, I set out to completely eliminate any visible support structure in the image and invented a system having an axial strut which is thin enough that its image fits within the reflection of the camera. The end of the strut closest to the camera is surrounded by a thick clear plastic window, which in turn is used in supporting the camera. This permits the strut to be "invisible" in the picture, even when used in the two-mirror "Cassegrain" wide angle reflector I made for panoramic imaging at the time. This system captures an unobstructed 360 degree panorama in a single picture. I use the shown simpler and larger axial strut reflector for imaging the night sky because it permits me to use a faster f/ratio.

An axial strut all sky or panoramic capture reflector can be built from relatively simple materials. I have issued and pending patents for some of my own improved axial strut reflector designs to reserve commercial rights, but individuals are typically granted a free license to construct and use a system such as the amateur axial strut hubcap reflector which is shown in the illustration for their own private noncommerical use, subject to the conditions set forth in the appendix of this book or in subsequent material I may publish on the subject.

2.3.2 Astrophotography From a Fixed Camera Mount

A simple camera on a tripod can capture constellations, star trails, meteors, conjunctions of the moon and planets, eclipses, and other interesting subjects. A wide field picture may not capture narrow angle objects as well as a telescope, but it can at least help you remember what you saw. For example, the lunar eclipse image in Figure 2.4 may not be of the best technical quality, but by looking at it, I can still recall the night I shot it. To get the picture, I just mounted my camera on a fixed tripod and took a 10 second exposure through a 105 mm lens. Only minimal streaking is visible in the 3× enlargement .

Photos taken during twilight or moonlight can allow you to use a relatively short exposure time and include silhouettes of terrestrial objects in your picture. Short duration tracking of the stars can be accomplished in a number of ways, including flexing your tripod; slowly panning a tilted tripod or side arm attachment (where one axis points toward the celestial pole); or using a slow motion head.

2.3.3 Piggyback Astrophotography

If you track the stars, impressive astrophotos can be shot with anything from an all-sky reflector to a telephoto camera lens. Long exposures are required for deep sky photography of nebulae and other dim objects,

Figure 2.4. Left: A 3 hour exposure on Ektachrome 200 through a 28 mm wide angle lens used at f/8 captures star trails which show Orion and Venus setting near Estes Park, Colorado. The lens was opened to f/3.5 for the last few minutes of the exposure in order to provide a bright spot at the lower end of each streak. Lights of passing cars caused the horizontal streak at the bottom. Right: A mere 10 second exposure through a 105 mm f/2.5 camera lens on ISO 200 color negative film was enough to image the Hyades star cluster and the relatively bright total lunar eclipse of 28 November, 1993 from light polluted Pasadena, California. This 3× enlargement encompasses about half of the original picture's field of view.

and proper tracking allows you to shoot such exposures without getting streaked images. Tracking during the time required for deep sky exposures can be accomplished with a commercial or home made tracking mount, or by piggybacking your camera on a telescope having a sidereal drive.

Piggyback photography is just as simple as its name. Your camera is literally piggybacked on your telescope. This relatively simple type of photography allows you to utilize the tracking capability of your telescope mount (or home made tracker) to get impressive photos of conjunctions, constellations, larger deep sky objects (nebulae), and other features of the night sky. In piggyback photography, you can utilize a wide range of camera lenses; from wide angle to telephoto. If your mounting is rigid enough, you can even "piggyback" a second telescope.

Most motorized telescope mounts can track well enough to take unguided photos of moderate duration with a wide angle or normal lens, but when a telephoto lens is used, it is a good idea to keep tabs on how accurately the mount is tracking and make any necessary adjustments. If your telescope mount has manual or motorized slow motion controls, you can correct tracking errors with relative ease. If it does not, you can try techniques such as slightly flexing the mount's tripod legs to achieve modest results.

When both your telescope and camera are rigidly mounted together, you can use your telescope to verify the tracking accuracy. Ideally, you would center a star in an eyepiece having an illuminated reticle (typically a cross-hair) and verify that the star image stays on the cross-hair during your exposure.

If you do not have a telescope mount with a motor drive or an illuminated reticle eyepiece, there are other tracking gadgets and guiding techniques which can be used, as will be covered in later sections. You can get

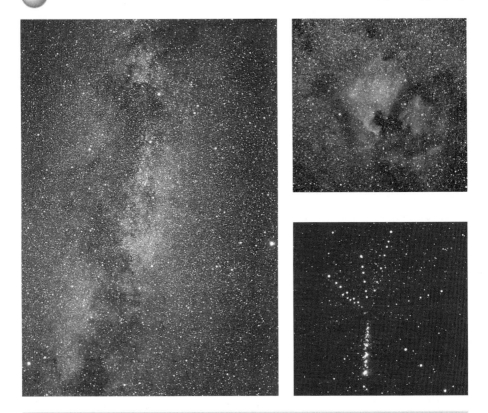

Figure 2.5. *Left:* This 10 minute exposure of Cygnus on ISO 1600 film through a 35 mm f/2 lens is reminiscent of the visual appearance of the same part of the sky from a dark site. The North American nebula is faintly visible near the top. Smaller objects such as the Dumbbell nebula (M27) are visible in the original image. *Upper right:* A fast f/ratio telephoto lens combined with fast film can capture dramatic images of dim deep sky objects in only a few minutes, but resolution and color saturation may be less than what you would get in a longer exposure with slower optics and film. This image of the North American nebula is a 6 minute exposure on ISO 1600 color negative film through a surplus Chicago Optical 200 mm f/2 lens which was stopped down to f/2.2. *Lower right:* A zoom lens can add interest to familiar objects. Here, a multiple exposure through a Tokina 50–250 mm f/4–5.6 zoom lens makes the belt and sword of Orion look like a scene from a sci-fi movie. The six overlapped one minute exposures on ISO 1000 slide film were taken at focal length settings of 50, 70, 100, 135, 170, and 210 mm.

started with very little financial cost, particularly if you make some items yourself.

Schmidt cameras seldom accept internal guiding attachments, so they are also applicable to piggyback photography. Most moderate sized units have a focal length in the 200–500 mm range. This is not unlike the focal length of moderately long telephoto camera lenses, but the Schmidt camera offers the advantage of a very fast f/ratio. This shortens the required exposure time and makes it possible to get truly extraordinary images of objects such as larger nebulae and comets.

2.3.4 Basic Astrophotography Through a Telescope

With the right accessories, a good telescope can usually be used as a long focal length camera lens which will permit you to photograph a vast array of terrestrial and astronomical subjects. Ideally, the telescope will have an eyepiece which is removable in order to allow a camera to be used in its place. A flip mirror is even better, since it will keep you from having to juggle your eyepiece and camera.

Figure 2.6. The full moon rises behind a saddle in the mountains near Estes Park, Colorado. This photo was taken through a 10 cm f/8 refractor, in the case, a Soligor camera lens. The exposure was about one f/stop less than what my light meter indicated for the surrounding sky.

Most telescopes have a focal length of at least a few hundred millimeters. A focal length of around 600 mm or more will provide useful photographic images of lunar phases. Twice as much focal length will begin to reveal details such as individual lunar craters. Fortunately, the moon is bright enough that it can usually be photographed without an equatorial mount or sidereal drive. With most telescopes, it can be relatively easy to get interesting photos of lunar phases, a moon rise, a partial lunar eclipse, a conjunction, or an occultation of another object by the moon. A sidereal

drive will become necessary if you want to shoot close ups of lunar craters or capture dim features such as earthshine on the crescent moon. With a proper solar filter, you can also photograph the sun.

2.3.5 Astrophotography With Auxiliary Optics

Many celestial objects have such a small angular size that a small to moderate size telescope will not provide a large enough image all by itself. This is where auxiliary optics come in. The effective focal length of a telescope can be increased with an auxiliary optic such as a Barlow lens or photographic teleconverter, but it is important to point out that a sidereal drive (which is used to track celestial objects) will become increasingly important as the effective focal length (and resulting image size) are increased. Effective focal length can be decreased with a telecompressor if the telescope has enough back focus.

The term "effective focal length" is used in association with auxiliary optics because the prime (or direct) focal length of the telescope itself does not usually change when you add an auxiliary optic; however, when a telescope is *combined with* an auxiliary optic, the combination provides a different overall focal length.

A more extreme increase in effective focal length can be accomplished by using the telescope eyepiece to act as an enlarging lens for the image. This technique is commonly referred to as "eyepiece projection". Eyepiece projection usually works best with an eyepiece of mod-

Figure 2.7. Center: The first quarter moon, photographed at a focal length of 2032 mm with a 20.3 cm f/10 Schmidt-Cassegrain telescope. Left: A 0.5× telecompressor provides an effective focal length of about 1000 mm with the same telescope. Right: A 2× Barlow lens provides an effective focal length of about 4000 mm.

erate complexity such as a Kellner, Orthoscopic or Plossl eyepiece. Surprisingly, a more complex eyepiece such as an Erfle or Nagler may not work as well. Too simple an eyepiece (such as a Ramsden) may not work very well either.

Another method of increasing the effective focal length is to use the "afocal" method, where you just point your camera and lens into your eyepiece. There are many advantages to this method, including the fact that you can use either a film camera or a video camera having a permanently attached lens.

Figure 2.8. Left: Central section of the first quarter moon photographed with a Questar 8.9 cm f/16 Maksutov–Cassegrain telescope, using a 16 mm Brandon eyepiece for projection to provide an effective focal length of about 6050 mm and an f/ratio of f/68. Upper right: Eyepiece projection was used with a filtered 10.2 cm f/15 refractor to photograph this sunspot near the solar limb at an effective focal length of about 30 meters. This image is a 2× enlargement from the original slide. Pincushion distortion from the projection eyepiece makes the solar limb look less curved than it should be at this image scale. Lower right: Jupiter, exposed for 6 seconds on ISO 1600 color negative film using the afocal method with the same Cassegrain telescope that was used for the left lunar photo. A 50–250 mm zoom lens was pointed into a 16 mm eyepiece which was used with a Barlow lens, providing an effective focal length of 32,000 mm. This image is only a 2× enlargement from the original negative.

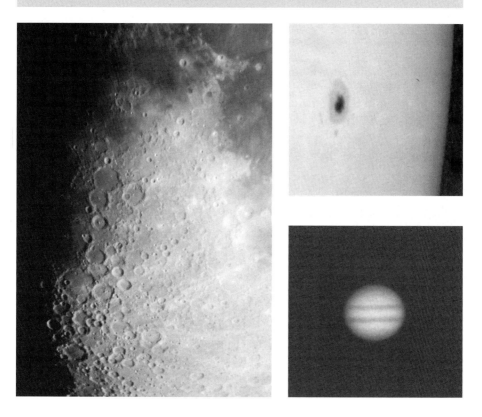

The sun can also be photographed through auxiliary optics, but only if a proper neutral density solar filter is used *in front* of the telescope. Solar filters which are placed entirely *behind* a telescope are unsafe; very unsafe. I have personally tested some of these behind the telescope filters on their respective telescopes (I don't look through the telescope during the test), and some of these filters have cracked or shattered in less than two minutes! If you are looking through a telescope when such a filter fails, your eye could be severely injured! You may even permanently lose all sight in the affected eye. In regard to attempts to look at the sun, the following precautions are in order:

NEVER try to center the sun by looking at it down the side of your telescope tube or camera lens; instead, look at the shadow of your telescope to see if it is pointing toward the sun. NEVER point any camera, telescope or finder scope toward the sun unless it is capped or has a PROPER solar filter installed in front of the objective.

NEVER look into (or place skin or other flammable objects just behind) a camera, binocular, or telescope that is pointed at the sun unless it has a proper front solar filter. NEVER leave your telescope or camera unattended when it is pointed at or near the sun, or when a bystander would be able to point it at the sun. If you must leave your telescope, cap it and point it well to the EAST of the sun. A single glimpse of the sun through an unfiltered or inadequately filtered optical system may result in INSTANT AND PERMANENT EYE DAMAGE WHICH MAY INCLUDE TOTAL BLINDNESS – even if you do not feel pain at the time your eyes are injured.

2.3.6 Deep Sky Astrophotography Through a Telescope

Guided deep sky photography is the most demanding type of astrophotography, but it is also the way you get those wowie zowie shots of galaxies, emission nebulae and other exotic objects. These photographs require exposure times which can vary from a few minutes to several hours. The effective f/ratio of your telescope will have a direct impact on how long an exposure will be required to photograph a given object with a given film. A larger aperture telescope will provide a larger image for an equivalent film, f/ratio, and exposure time. Telecompressors are available for some telescopes. These provide a faster f/ratio at the expense of a reduced image size and possibly some vignetting.

The combination of a long exposure time and a long focal length impose significant demands on the tracking accuracy a telescope mounting must have. In fact, the tracking accuracy required for a good image can easily exceed the mechanical accuracy of even the best commercial telescope drive system. As a result, one usually cannot get a good deep sky photo through a telescope if the drive is just left to run without some sort of feedback from either the astrophotographer or an autoguider. I say "usually" because there are a few unique (though limited) exceptions to the rule. The most common result of uncorrected tracking is a streaked image.

In order to properly guide during your photograph, you will need the appropriate accessories for your telescope. For this reason, guided astrophotography through a telescope is relatively demanding in regard to minimum equipment requirements. Most astrophotographers utilize an off-axis guider or separate guide scope. Other guiding systems include a beam splitter, a partial aperture guiding reflector (which has a reflector small enough that it only intercepts part of the light cone from the telescope), or a multiple function image switching attachment (such as the patented Versacorp DiaGuider) which doubles as an off-axis guider.

Widening the guide star selection area is a very useful feature. Some off-axis guiders have radial adjustment, accomplished by sliding the eyepiece over a fixed reflector (as in the patented Lumicon Easy Guider), sliding the reflector in or out (as in the patented Versacorp DiaGuider and VersaGuider), or tilting the guiding reflector (as in the original prototype of the Versacorp VersAgonal and more recent products such as the Celestron radial guider). Other adjustable guiders are based on different concepts.

In addition to the basic guiding assembly, you will need some sort of drive corrector to modify the drive rate of your telescope mount (some mounts have built in drive correctors), a power source for the telescope drive, and a guiding eyepiece or autoguider. Some of these guiding accessories are not cheap, so not all astrophotographers get into guided deep sky photography right off the bat.

Guiding during your photograph can be tedious, but it is not all that bad if your guider has a manual shutter or some other means of interrupting the exposure without cutting off the guide star image in your eyepiece. Such a shutter feature will allow you to take a break as often as you like without spoiling your picture. This is handy when a friend yells "Wow, look at what I've got in my 45 cm telescope!" My guider has a manual shutter, so all I have to do when that happens is close the manual shutter, go over to my friend's big telescope, and take in a view that knocks my socks off!

The wide variety of available astrophotography equipment can satisfy even the most enthusiastic gadgeteer, and more and more aspects of astrophotography are becoming increasingly simple. It is getting to the point where even a trained monkey can take astrophotos; however, it is possible to get so many gadgets that you end up spending more time setting them up than you do observing or taking pictures. This is fine if you like setting up gadgets, and it can be necessary in photographing short lived events; but moderation in gadgetry can be the best policy for most astrophotography.

Many astrophotographers now use autoguiders on a regular basis and take naps during their exposures. Taking photos in one's sleep can be nice, but some autoguider units can cost as much as a good telescope. It can also take time to set up an autoguider for each picture (particularly the first one on a given night), so such a unit may offer only a marginal advantage over manual guiding for shorter exposures. An autoguider can be handy for really long exposures, but it is by no means a necessity.

Figure 2.9. Left: Front view of an early off-axis guider with a fixed reflector. The off-axis reflector position permits guiding on an off-axis star without obscuring the center of the picture. A reticle eyepiece or autoguider is used in the top of the unit and the camera is mounted behind. Right: First introduced in 1987, the patented Versacorp Deluxe DiaGuider (TM) (one of my own inventions) combines the functions of a star diagonal, flip mirror, and adjustable off-axis guider. These front views show the DiaGuider's large multiple function reflector in its axial subject acquisition and star diagonal position (left); and one of its off-axis guiding positions (right). The small knob low on the left side controls a built-in manual shutter plate. The optional MicroStar (TM) focusing attachment (not shown) can be used on top of the DiaGuider without disturbing a rear mounted camera.

There is no point in putting off deep sky astrophotography until you get an autoguider. I did not use one for any of my own photos in this book; and if I can get good astrophotos without an autoguider, you can too. Many clear nights will slip through your fingers if you just sit around and wait until you get a given astro gadget before you go out and observe or take pictures. The camera or telescope equipment you already have may be enough to at least allow you to get started in some type of astrophotography.

2.3.7 Event Photography

Observing astronomical events such as conjunctions, occultations and eclipses can be truly memorable. Photographing such events usually requires at least a few special considerations. An astronomical event occurs only at its set time. It does not wait for anyone, so the astrophotographer has to do the job right the first time. Unlike conventional astrophotography, you don't have a second chance to shoot the same subject at a later time. In the case of a total solar eclipse, you only have a few short minutes to observe it and take your photos; then it's all over.

Preparing for astronomical events provides a good opportunity to evaluate and refine your equipment. It is always a good idea to set up your equipment and practice your procedures at least a day before each of your first few local events. It is also a good idea to prepare a checklist for your equipment and photographic procedures. Taking time to do this will allow you to shake down your equipment setup and add a few features or items you may need.

It is not unusual for one to want to both observe and photograph an astronomical event. Most events don't last very long, so juggling an eyepiece and camera do not make for a good time. This explains the increasing popularity of amateur and commercial flip mirror attachments. These allow a camera and eyepiece to be simultaneously attached to a telescope.

Repeated practice is very useful even if you are not photographing events. Among other things, it can help you identify small improvements you can make to your equipment. Some of these improvements may be easy to make. If you make any improvements, it is a good idea to test your improved equipment before you actually use it in the field. Repeatedly practicing your setup

and procedures in a familiar area can help you comfortably run through your procedures several times and identify features or methods which may need improvement. Occasional practice in unfamiliar surroundings can also be useful.

2.3.8 Astrophotography for Research

Many people take astrophotos for aesthetic purposes, but astrophotography is also useful in scientific research. Research applications do not necessarily involve different techniques in taking the actual image, but the emphasis or use of the image may be different. In research, it is very important to accurately record data such as the exposure time, film type, observing conditions, and universal or local time.

In most cases, it is best to use universal time or U.T. (as opposed to local time) when reporting observations because universal time is the same everywhere on earth and is not subject to daylight savings time or other whims of local politicians. Another term often used for universal time is GMT, for Greenwich mean time, since universal time is the same as local standard time in Greenwich, England, which is at a longitude of 0 degrees.

A few shortwave radio stations such as WWV and WWVH broadcast coordinated universal time at frequencies of 2.5, 5, 10, and 15 MHz. The broadcast typically indicates the beginning of each minute with a one second tone, followed by shorter tones every second.

Figure 2.10. Left: My first astrophotography setup did not even include a tripod! Rather than miss out on many clear nights, I used a typing table to support my 9 cm f/11 Maksutov Cassegrain telescope and its sidereal drive. Right: Earthshine is visible on the young crescent moon a few minutes before it occults Jupiter and its moons on the evening of 15 July, 1980. This 15 second exposure at f/11 on Ektachrome 200 film was taken with the setup at left. The image is enlarged about 4×.

Near the end of each minute, the tones are replaced by clicks and the time for the upcoming minute is stated in the English language. This is followed a few seconds later by the tone for the start of the next minute.

Organizations such as the International Occultation Timing Association (IOTA) assist in coordinating and reporting observations as well as publishing good material on how to effectively observe events. Amateurs and professionals alike can thereby contribute to meaningful research.

There are individuals and groups of various financial means who devote virtually all of their efforts to astronomical research and public education. Some groups have shown that serious research can be performed with modest material resources. A good example of such a group is Astronomia Sigma Octante (ASO); a center for astronomical research in Cochabamba, Bolivia. I had the pleasure of meeting several ASO members while in Bolivia for the 3 November 1994 total solar eclipse. Even though ASO has relatively few members and comparatively modest equipment, the extent of their photos and observation records (particularly those of the director, German Morales) are quite impressive. In regard to real science, their accumulated work easily exceeds that of some more generously funded organizations back here in North America.

Astrophotography for research can apply to observations of the sun, variable stars, eclipse and occultation timings (including occultations by asteroids) meteor showers, comets, or even a nova or supernova. Some observations are best conducted visually or with video, while others can benefit from astro imaging with a film camera or with an integrating electronic imager.

Solar photography with a solar filter of adequate density can supplement visual sunspot or eclipse observations. Astrophotography also provides good data for use in identifying star fields related to variable star or comet observations and as a reference in preparing related drawings or reports.

Comets are interesting and sometimes beautiful objects which are photographed for both aesthetic and scientific purposes, and cometary photography can provide rewarding images. Amateur and professional photographs have been used in research to both discover comets and to refine their orbital data.

It is not unusual for a comet to have a relatively rapid proper motion in relation to the background stars, so appropriate special guiding techniques may be

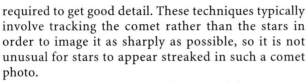

required to get good detail. These techniques typically involve tracking the comet rather than the stars in order to image it as sharply as possible, so it is not unusual for stars to appear streaked in such a comet photo.

Many comets have been discovered by amateur astronomers, with visual observation resulting in the most amateur discoveries. It is unlikely that your astrophotos will result in the discovery of a comet, but there are sill occasions when amateur photos can be useful for research. Sometimes, after a comet is discovered, it is later located on photographs which were taken prior to its discovery, and such photos (if well documented in regard to when they were taken) can be just as valuable in refining orbital data as those taken after discovery. Well documented photographs of known comets are also useful in keeping their orbital data up to date.

Occultations are some of the most interesting and dynamic events to photograph, and well documented observations can provide valuable data. It is here that we can refine lunar limb profile and lunar position data, discover the shape and size of an asteroid, or even discover that a star which is single on the charts is actually a double star. Occultations can typically be accurately predicted, and continued observations can in turn make it possible to predict future events with even greater precision.

Many occultation measurements are now made by amateur astronomers, and it can be fun to participate in group efforts to observe and document such events. Some of these efforts concentrate on grazing occultations by the moon or occultations of stars by asteroids. Here, a group of amateur astronomers will typically set up along available roads which run more or less perpendicular to the path in which the event is observable. Where such a linear array of observation sites is not possible, people just observe from where they can within the event path; then their relative positions are taken into account when the data is reduced.

The size of an asteroid can be determined with some precision by measuring the width of its "shadow" when it temporarily covers a star. The outline of the shadow can be determined by compiling the observations of several observers and correcting for the position of each one.

The width of the asteroid shadow is determined by documenting which observers witnessed an occultation and which ones did not. The other visible dimension of the asteroid (few asteroids are perfectly spherical) can be determined by the amount of time the star is covered. This is possible because the asteroid's shadow typically moves at a known speed in relation to the earth's surface.

By compensating for the difference in time between each of the observations, drawing lines which represent the distance traveled by the shadow while the star was occulted during each observation, drawing a smooth shape which intersects with the ends of the lines (so each line forms a chord), then taking into account the incident angle of the shadow to the earth's surface, the approximate profile of an asteroid can be determined.

In some occultation events, one can see an asteroid approach a star over a period of minutes, then the star may appear to dim for a few seconds and brighten again. In events where the asteroid is so small and dim that it can't be seen in an amateur telescope, the star may just appear to briefly dim or go out, then brighten again.

In the absence of an occultation event, the size and shape of an asteroid must be estimated from its brightness or other less precise methods. Asteroids are very small in terms of angular size unless they make a very close (and very rare) approach to Earth. In fact, the angular size of a typical asteroid is so small that it will only look like a small dot or a short streak on a film image, depending on how much its orbital motion causes it to move in relation to the background stars during the exposure. Asteroids have rarely been imaged in detail, and when they have been imaged from earth, some of the best images have been obtained with large antenna arrays working at wavelengths substantially longer than what "optical" telescopes utilize.

Audio taped comments or video taped event observations are the most useful because they allow for a more precise reduction of data. Here, the observer can watch or listen to a recorded observation multiple times. In my own observations, I use a metal party clicker to produce a sharp sound whenever each stage of an event occurs, then begin to describe what happened on tape. This allows me to record the time of the event without the inconsistent delay that could result from trying to mentally compose a description for each event before indicating that something happened. This in turn results in a shorter reaction time, which is often called a "personal equation" in data reduction.

Similar observing techniques can be used to document a grazing occultation by the moon. At some rare grazing occultation events, one may see a star wink on and off up to several times as it is covered and uncovered by mountains on the edge of the moon. In other events, the star may not entirely or suddenly disappear all at once, indicating that it may be a close double star. Grazing occultations can be the most interesting, but it is also useful to observe lunar occultations from sites which will not experience a grazing event. When well documented, all of these observations will facilitate more precise predictions of the lunar limb profile, which in turn make it easier to predict the nature of future occultations or even some aspects of solar eclipses, particularly in cases where eclipses are observed from near the limit of totality or annularity.

Novae and supernovae are events which can occur without advance warning. Some are discovered by direct visual observation, while others have been discovered by amateur or professional astronomers who simply notice something is out of the ordinary on their photographs.

Figure 2.11. Now you see it, now you don't! These off-axis guided photos show the galaxy NGC 5128, with (left, arrowed) and without (right) a supernova in the field. The left photo is a one hour exposure on Konica 1600 color negative film with a 20 cm f/10 SCT and a Versacorp VersAgonal with a built-in 0.7× telecompressor. The right image is a 50 minute exposure on ISO 1000 3M slide film with the same optics. Both images are enlarged about 4×.

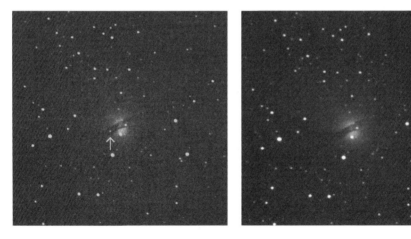

Chapter 3

Calculating Focal Length, f/ratio, etc.

Astrophotography covers a wide variety of disciplines. Sometimes, it is necessary to perform a few calculations in order to determine the effective focal length and f/ratio of your telescope. Such calculations can permit you to more accurately determine what exposure time you should use, which in turn will save film. A little math is particularly important when auxiliary optics are used to change the effective focal length of your telescope. Shooting an astrophoto without first knowing the correct exposure (or at least something close to the correct exposure) is literally like taking a "shot in the dark".

Fortunately, the necessary calculations are not very difficult and you can often get by without having to come up with an exact answer; just being within ten percent or so is acceptable for most applications. To make things easier, some equations in this chapter are presented in a relatively illustrative (and sometimes unconventional) fashion in order to better explain their relevance to real world astrophotography and make them easier to grasp for those who may not have done much math lately. The average astrophotography related equation is so simple that it can typically be solved in seconds rather than minutes. It's easy as "pi"!

One of the most important things you can do is document your exposure times and the way you set up your equipment. This will allow you to use your existing pictures and exposure data as a reference when taking future photos. Sometimes, it can also be helpful to make a simple drawing of your subject and/or parts of your equipment. A sample form for recording exposure data is provided in the appendix of this book.

x

This chapter will stress calculating your exposure value in a step by step manner by using a series of simple equations rather than using more compact and conventional mathematical expressions. I have found this step by step method more useful because it allows me to calculate most things in my head while using my telescope at a dark site. This in turn preserves my dark adaptation because I usually don't have to use a light, calculator or writing instrument while making my calculations.

Figure 3.1 is a good example of how to "wing it" in unfamiliar situations and still get properly exposed pictures. While visiting Kitt Peak observatory in the 1980s with the Saguaro Astronomy Club of Phoenix, Arizona, I had the unexpected opportunity to briefly look through a 0.9 meter telescope for the first time. Saturn was the first planet we got to see, and the view of it through such a large aperture telescope was tremendous! I wanted to grab a picture if possible, but each person in line had only about a minute at the eyepiece. This left no time to scribble on paper while calculating the required exposure.

The first step was to figure out the telescope's magnification by comparing its image of Saturn to something of a known angular size. Fortunately, I recalled that the finger nail on my index finger occupies an angle of just over a degree (or about two lunar diameters) when viewed at arm's length. The average angular diameter of Saturn's disk is about 1/100 that of the moon, which made it 1/200 that of my finger nail. Since the image of Saturn in the eyepiece looked about twice as large as my finger nail, the magnification of the telescope had to be about 400×.

To shoot the picture, I used the afocal technique, which is just pointing my camera and lens (hand held

Figure 3.1. Knowing how to quickly estimate effective focal length, f/ratio, and required exposure time in my head allowed me to quickly grab these afocal images of Saturn and Mars through a 0.9 meter Cassegrain telescope at Kitt Peak observatory. The vertical line through Saturn is from a reticle in the telescope eyepiece. These pictures are printed backwards in order to compensate for left to right image reversal from the telescope's right-angle eyepiece holder.

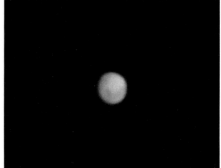

in this case) into the telescope eyepiece. Calculating the focal length was easy; with the afocal method, the focal length of my 50 mm camera lens is increased by a factor equal to the telescope's magnification of 400×; providing a focal length of 20,000 mm. The f/ratio of about 22 was calculated by dividing the 20,000 mm focal length by the 900 mm aperture of the telescope. At the time, I had developed the beginnings of an exposure guide like the one in the appendices of this book from documenting my exposure times, so I knew the correct exposure for Saturn on Ektachrome 200 film at an f/ratio of 22 was about 1 second. This 4× enlargement speaks for itself, though it would have been better without the vertical black line from a reticle in the telescope eyepiece.

Later the same evening, I had the opportunity to photograph Mars in a gibbous phase through the same telescope. Mars is brighter because it is closer to the sun, so it only required an exposure of 1/15 second. I have since had the opportunity to shoot many other pictures through big telescopes on public viewing occasions. This would not have been possible if I had not known how to rapidly determine the visual magnification and calculate the correct focal length, f/ratio and exposure.

This chapter will show that most astrophotography calculations do not require much in the way of math skills. In addition, you don't have to calculate everything exactly in order to get good results. For example, when I measured the size of my Saturn photo and compared it with the angular size of Saturn on the date I took the pictures, I learned that the effective focal length was really about 10 percent longer than what I had estimated, yet the picture came out fine! Knowing how to estimate the approximate required exposure can allow you to get good shots through any good telescope you have access to, even if you don't have a telescope of your own.

3.1 Visual Magnification

Visual magnification (sometimes informally called "power" by amateur astronomers) is an absolute value which is determined by dividing the focal length of a telescope objective by the focal length of its eyepiece.

Let's say that a telescope has a focal length of 2000 mm and it is used with a 25 mm eyepiece. The visual magnification is: 2000/25, or 80×. When you look through the telescope, the image of a given object will literally look 80 times larger than it does from the same vantage point with your unaided eye. This is easy to visualize when you consider how magnification is accomplished. In essence, a 25 mm eyepiece is a magnifying lens of 25 mm focal length. If you were to use the optics from your eyepiece as a loupe, you would be able to observe small objects in detail. When looking through the eyepiece, you are able to look at the image of a subject from an optical distance of 25 mm.

When you use the eyepiece with a telescope, you see the image formed by the telescope's 2000 mm focal length optical system from an effective distance of only 25 mm. Since the telescope's focal length is 80 times longer than that of the eyepiece, the subject looks 80 times larger. A shorter focal length eyepiece will provide even more magnification because it allows you to look at the telescope image from a closer optical distance. The visual angle of view at a given magnification is a relative value because it is dependent on the size and resulting apparent angular diameter of the eyepiece field stop (or the "apparent field" of the eyepiece); different eyepieces of the same focal length do not necessarily have field stops of the same size.

A lot of interesting interrelations can be used to provide shortcuts for some astrophotography calculations. Some of the interrelated aspects include telescope focal length, aperture, f/ratio, eyepiece focal length, and exit pupil diameter. (The exit pupil is an image of the telescope objective entrance pupil which is formed by the eyepiece.) Some interrelated ratios are shown below:

$$\frac{200 \text{ mm telescope focal length}}{25 \text{ mm eyepiece}} = 80\times \text{ magnification}$$

$$\frac{25 \text{ mm eyepiece focal length}}{\text{Focal ratio of } 10} = 2.5 \text{ mm exit pupil}$$

$$\frac{200 \text{ mm aperture}}{80\times \text{ magnification}} = 2.5 \text{ mm exit pupil.}$$

Note how the exit pupil diameter can be determined by either the eyepiece focal length and the f/ratio, or by the telescope aperture and the magnification. The exit pupil diameter is of particular interest in eyepiece projection and afocal photography, since it will allow you to estimate the working f/ratio (the value you need to know in calculating your exposure time) without even having to know the effective focal length of your setup.

3.2 Focal Length and Photographic "Magnification"

Magnification is often discussed in the field of astrophotography, but it is not a very appropriate term for photographic images. In astrophotography, an image is formed on film. As a result, magnification is entirely relative, since it depends on the distance from which the final image is observed.

Often, relative photographic magnification is specified in terms of the ratio of the telescope focal length to that of a normal camera lens, but even this can be ambiguous. For instance, a 35 mm camera may utilize a normal lens of 50 mm focal length, while a normal lens having the same field of view for a 6 × 7 cm format camera would have a focal length of about 100 mm. Thus, a telescope of 2000 mm focal length would have a relative magnification of 40× (2,000/50) for the 35 mm camera, but only 20× (2000/100) for the 6 × 7 cm camera. To avoid this type of confusion, photographic images are typically specified in terms of an absolute value such as focal length, image scale, or angle of view. Of these, focal length is most commonly used in amateur circles.

Focal length is the optical distance between the optical center of an optical system and the center of the focal surface. Focal length is easy to envision when you think of your optics in terms of a pinhole or simple positive (convex) lens which is placed a given distance from the focal plane. In this situation, the distance from the pinhole or lens to the focal plane will be equivalent to your focal length.

To evaluate the field of view with a telescope, just envision two straight lines converging on a pinhole from the outer angular extremes of a distant subject. When these straight lines intersect at the pinhole and continue through it, they will diverge behind it at the

same angle which they had converged in front. From this, it is easy to see that the image will be larger if the effective focal length is made longer by moving the focal plane farther from the pinhole; the more distance between the pinhole and the focal plane, the wider the lines can diverge before they reach the focal plane. This will obviously result in a larger image scale.

Figure 3.2. Left: Laguna Beach, photographed with a 35 mm camera and normal 50 mm lens. Right: Gull photographed from the same location, but through the long focal length telescope and Barlow lens shown in the lower right of the previous image. The focal length of 2500 mm provides a vertical angle of view of about half a degree; roughly the same angle occupied by the moon. At this focal length, the gull is imaged 50 times larger than it is in the left photo, where it appears to be a mere dot under the arrow.

Some types of astrophotography require a focal length far longer than that used in ordinary terrestrial photography. For example, let's say you want to get a picture of the full moon which will fit comfortably within the 35 mm format. The actual size of the 35 mm format is 24 × 36 mm, but you may want to limit the lunar image to about 20 mm in order to allow room for cropping and a moderate degree of pointing error. When the moon is at perigee (the closest point in its orbit to earth), it is about 100 times its own diameter away from us. Therefore, the focal length required to obtain a given image size will be about 100 times the desired diameter of the image, so to get a 20 mm diameter image of the moon, you will need a telescope of 2000 mm focal length. (There can be some variation in how large a lunar image you will actually get because the moon's elliptical orbit causes its angular size as seen from earth to vary by more than 10 percent.) 2000 mm is substantially longer than even the "super telephoto" camera lenses used for terrestrial photography.

If you want to photograph a planet and get a reasonable amount of detail, you will need even more focal length. Many popular deep sky objects also have an angular size smaller than that of the moon, though not usually as small as that of a planet. For these (with the exception of a few larger "showpiece objects"), you usually have to settle for less than a full frame image.

Figure 3.3. Left: Lunar images obtained at different focal lengths are shown relative to the 35 mm format in this composite picture. The image boundary in this 1.8× enlargement corresponds to the portion of the 24 × 36 mm format typically utilized in prints made by commercial photo labs. From left to right are images shot at focal lengths of 100 mm, 200 mm, 400 mm, 800 mm, and 1600 mm. Right: Typical images sizes for Jupiter at opposition are shown here. From left to right are relative images sizes obtained at 2000 mm, 4000 mm, 8000 mm, 16,000 mm, and 32,000 mm. The angular size of Jupiter can be more than 30 percent smaller when it and the earth are on opposite sides of the sun.

3.3 Image Scale and Field of View

For a given film format or eyepiece field stop, image scale directly influences field of view; when one goes up, the other must come down. More particularly, if image scale is increased, the image of the subject will be larger but the field of view will be decreased, or vice versa.

If the film format or eyepiece field stop size is changed, then the relationship between image scale and field of view becomes more relative. For instance, a 20 mm focal length wide field Nagler eyepiece will cover a wider field than a typical 28 mm orthoscopic eyepiece, even though the Nagler eyepiece provides almost 50 percent more magnification.

Some astrophotographers specify their photographs according to field of view, then compare this with the field of view for a given eyepiece and magnification. In doing this, it is important to remember that magnification of the photographic image and the angle of view for the eyepiece field may both only be relative values, while the angle of view for the photographic image and the magnification of the visual image in the eyepiece would both be absolute values.

In some research applications, it is useful to specify an image in terms of angular image scale. This value can be calculated when one knows the focal length used to acquire the image. To get the field of view in degrees per millimeter, just divide 57.3 by the telescope focal length in millimeters, then multiply the answer by the width of your film format. For example, a 2000 mm telescope will provide an image scale of 57.3/2000, or 0.02865 degrees per millimeter. Therefore, the field of view across the long dimension of a 24 × 36 mm format is 36 mm multiplied by 0.02865, or 1.03 degrees.

To get the field of view in arc minutes, substitute 3438 for the 57.3 figure (or multiply 57.3 by 60), and to get the field of view in arc seconds, simply substitute 206,265. These figures are the ratio of the field of view for the angular unit of measure to the distance from the observation point. For example, a one degree field of view would cover a one meter area from a distance of 57.3 meters.

Field of view can also be calculated in other ways. Trigonometric functions are particularly advantageous for such calculations, and electronic calculators make this just about as easy as any other method. It is usually helpful to calculate the approximate field of view for all of your optics in advance of your observing or astrophotography sessions. This will make it easier for you to know which optics will provide the results you want.

To most accurately calculate the angle of view when you know the focal length of an optical system, it is best to envision a right triangle which matches specific locations on the film format (or eyepiece field) and in the optics. The intersection of the optical axis and the focal plane forms an angle of 90 degrees, which can be the 90 degree corner of our right triangle. (This intersection also corresponds to the center of your picture.) The other corners of the triangle are at the edge of the format and at the optical center of your optical system (which is not always the physical longitudinal center of the optical system). This triangle defines the angle of view on only one side of the optical axis (only half of the format width is defined by a side of the triangle) so we must double the result of our calculation in order to get the angle of view across the full width of the format. This method works for optics which provide a rectilinear image projection, which is the projection most photographic optics (other than fisheye lenses) tend to have.

To calculate the angle of view that a normal 50 mm lens will provide across the long dimension of a 24 mm × 36 mm format, we start by taking the ratio of half the 36 mm format width dimension (18 mm) to the 50 mm focal length of the lens:

$$18/50 = 0.36.$$

From here, we calculate the inverse tangent of the ratio, then double the result:

> Inverse tangent 0.36 = 19.8 degrees;
> 19.8 × 2 = 39.6 degree angle of view.

To calculate the focal length when you know the field of view or the measured size of a subject having a known angular subtense, simply take the tangent of half the angle of view, then divide half the format width by the result:

39.6/2 = 19.8 degrees; Tangent 19.8 degrees = 0.36; 18 mm/0.36 = 50 mm focal length.

3.4 Determination of Your Telescope's Focal Length and f/ratio

Most telescopes list their focal length and f/ratio on a lens cell or a label located elsewhere. If they do not, you can roughly determine focal length, etc., by making a few measurements or performing a few tests. The testing methods in particular can be valuable, because they will allow you to determine the effective focal length of your telescope when using auxiliary optics. Some types of telescopes can be measured outright, but the type of telescope you have will determine the proper measurement method.

3.4.1 Calculating the Focal Length and f/ratio of a Refractor

Most refractors have a relatively straightforward design. If all of the optics are in the front cell of the telescope, you can get a rough idea of your focal length by simply measuring the distance from about the middle (longitudinal center) of the objective to the focal surface. (The term focal surface is used instead of focal plane because the surface of best focus for most telescopes is not perfectly flat.) This measurement method is not exact, but it is close enough for most applications. Some variation in accuracy can depend on whether your telescope is a doublet or triplet; and in the case of a doublet, whether a positive or negative lens is used in front. The focal length of some refractors cannot be measured so easily. For example, some Tele-Vue® refractors are an "aplanatic" design having an additional positive lens a substantial distance behind the front elements. This typically causes the physical length of the telescope to be considerably longer than its optical focal length. Fortunately, the focal length of such a commercial telescope is well labeled.

To get the f/ratio of a refractor telescope, just measure the clear diameter of the front element. Then, divide the focal length by the lens diameter, being sure to use the same units of measure for each value. For example, say the measured distance from the lens to the focal plane is 80 cm and the clear aperture is 10 cm. Your f/number will be 80/10, or 8, meaning that the aperture is $\frac{1}{8}$ of the focal length. The most common way to indicate f/ratio is to substitute the letter "f" for the number 1. Therefore, the telescope f/ratio would be "f/8".

A few refracting telescopes (mostly cheap ones) may have baffles which keep them from utilizing their entire front element, meaning that they have less useful aperture than one may expect. You can estimate how much of an objective is utilized by looking backwards into the edge of its objective. If you can see the center of the eyepiece aperture from this position, the baffles are not limiting the aperture at the center of the image. It is not usually a problem if the baffles restrict a modest part of the aperture for off-axis parts of the image.

3.4.2 Calculating the Focal Length and f/ratio of a Newtonian

Most Newtonian telescopes are relatively straightforward to measure. Just measuring the length of the tube can get you in the ball park, but if you want to be more exact, measure the distance from the center of the primary mirror to the center of the diagonal mirror, then measure the distance from the center of the diagonal mirror to the center of the focal surface where an object at infinity is imaged, then add the two distances together. Be careful not to scratch the mirrors while making your measurements.

You can also measure the focal length when the primary mirror is not in the telescope by measuring the distance from the front of the primary mirror to the position at which it forms a sharp image of a distant subject like the moon. (It is best not to use the sun because you could injure your eye by accidentally getting it into the light cone or even looking obliquely at an unfiltered solar image.) Another method of determining the focal length does not involve pointing the mirror at anything. Instead, look directly into the mirror with one eye and move toward or away from it until the pupil of your eye appears to fill the entire mirror surface, then divide the distance between your eye and the mirror by two.

To get your f/ratio, measure the primary mirror diameter or the aperture at the front of the tube (whichever is smaller), then divide the focal length by the measured diameter. Some telescopes having a Newtonian configuration also have correcting optics in front, or they may even have correcting or Barlow optics in the focuser tube. Certain types of these extra optics can significantly modify the focal length. For such telescopes, you will need to measure the focal length with techniques like those used for the Cassegrain telescope.

3.4.3 Calculating the Focal Length and f/ratio of a Cassegrain

The focal length of a Cassegrain telescope is not easily measured with a ruler or tape. Accordingly, other methods must be used. These methods can also be used with other types of telescopes. One easy way to adequately measure the focal length is to measure the image scale from a photo taken with the telescope. When the moon is at perigee (the closest point in its orbit to earth), it is about 100 times its own diameter away from us. At the most distant point in its orbit, the moon can appear more than 10 percent smaller. If an image of the moon at perigee is 20 mm in diameter, then it follows that the focal length will be about 100 times longer than the diameter of the image, or 2000 mm. A far off terrestrial object of a known size and distance can also be used for a test subject.

Other methods do not require that you take a picture. You can often get an adequate idea of a telescope's focal length just by comparing the relative size of the moon's image to the size an eyepiece field stop or known features in a camera viewfinder. For more accuracy, you can instead measure the image scale of a known object with a graduated reticle or simply time the drift of a star across a field of view having a known width. For the later method, turn off your sidereal drive, point the telescope just west of a star which is near the celestial equator, and measure the time it takes for the star to drift across a camera finder or eyepiece field having a known size. Rotation of the earth will cause the sun to drift one degree (about twice its own angular diameter) in exactly four minutes when it is at the celestial equator. A star on the celestial equator will drift one degree in about 3.989 minutes (239.3 seconds). Due to its orbital motion, the moon will drift at a slightly slower rate than either the sun or a star. In general, you can determine your focal length within a few percent if you just use the four minute figure for any object that is near the celestial equator.

You can start out by noting the time it takes for a star near the celestial equator to drift across the long dimension of your SLR camera finder. The full frame 35 mm format is 24 × 36 mm, but the focusing screen in most modern cameras is only about 33 mm wide. (You usually do not want to use the short dimension of the viewfinder because the SLR mirror may be so short from front to back that it cuts off part of the field.) Let's say that it takes a star near the celestial equator exactly three minutes to drift across the screen. This is $\frac{3}{4}$ of the time it would take to drift one degree; therefore, the field of view is $\frac{3}{4}$ of a degree wide.

In calculating your focal length, you can either reference your drift time to the four minute time, or calculate it entirely from scratch using trigonometric functions. I prefer the comparative approach because it is easier to do in my head when I'm out in the field. To get your field of view in degrees, just divide 57.3 (mentioned in Section 3.3) by the decimal value of the field of view in degrees: 57.3/0.75 = 70.7; then multiply the result by the width of the camera's focusing screen: 70.7 × 33 = 2333 mm focal length.

Trigonometric functions (see Section 3.3) are most often used to calculate the angle (or field) of view when the focal length is known, but the technique is also applicable to calculating focal length when you know the angle of view.

Some Cassegrain telescopes do not have a fixed focal length. For example, the effective focal length of an internally focusing Cassegrain telescope (specifically one which focuses by moving its primary or secondary mirror) can vary slightly according to how far behind it a camera is positioned. In general, the focal length will increase when the back focal distance is increased. This occurs because the primary and secondary mirrors must be moved closer together to accommodate the increased back focal distance. This in turn causes the axial light cone from the primary mirror to encompass a larger part of (and a greater degree of curvature on) the secondary mirror, thereby effectively increasing the secondary mirror's amplification factor. An extreme increase in back focal distance can have other effects, including a reduction in the telescope's effective aperture if the light cone exceeds the inner diameter of the telescope's primary light baffle tube. Changes in focal length versus back focus of an internally focusing Cassegrain can also influence the effective magnification of telecompressors and some other auxiliary optics.

3.4.4 Calculating the Focal Length and f/ratio of an Unusual Telescope

Off-axis telescopes including the Tri-Schiefspiegler and Yolo are not very common, but if you have one, you can measure the focal length using techniques like those for a Cassegrain telescope. The same goes for evaluating most other unusual telescopes. Some off-axis telescopes may have a tilted focal surface which will have to be accounted for when attaching a camera.

3.5 Calculating Effective Focal Length and f/ratio with Auxiliary Optics

An auxiliary optic such as a telecompressor or Barlow lens can modify the effective focal length of your telescope, but this will in turn change the working f/ratio of your system. Before you try to take astrophotos with

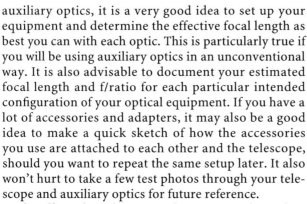

auxiliary optics, it is a very good idea to set up your equipment and determine the effective focal length as best you can with each optic. This is particularly true if you will be using auxiliary optics in an unconventional way. It is also advisable to document your estimated focal length and f/ratio for each particular intended configuration of your optical equipment. If you have a lot of accessories and adapters, it may also be a good idea to make a quick sketch of how the accessories you use are attached to each other and the telescope, should you want to repeat the same setup later. It also won't hurt to take a few test photos through your telescope and auxiliary optics for future reference.

The effect of nearly any auxiliary optic is dependent on the distance between it and the focal surface. In addition, the effective focal length of some telescopes can actually change as they are focused. This is the case with most internally focusing Cassegrain telescopes; in these, the effective focal length usually increases by at least three times as much as any increase in its own back focus distance. If you want an exact answer, this focal length change must be accounted for in your other calculations.

Even in conventional configurations, not all auxiliary optics will perform exactly as specified. I have seen 2× Barlow lenses which really operate at 1.8×, and 0.5× photographic telecompressors that really work at more like 0.6×. Some eyepieces may differ by five percent or more from their indicated focal lengths. All of these variations are cumulative in your setup, so measuring your effective focal length (at least for your most often used configurations) can be a good idea. To do this, you can use the same techniques used for the Cassegrain telescope in Section 3.4.3.

The required quality of an auxiliary optic depends on a lot of factors, but it is safe to say that the quality must be highest if the optic is used with a telescope having a fast f/ratio. A fast f/ratio telescope will reveal defects that would not be detectable if the same auxiliary optic were used on a slower telescope. This is because a greater diameter of the auxiliary optic is used in imaging any given point of the subject if the telescope has a faster f/ratio. In general, you can get good results with relatively mediocre auxiliary optics if the telescope is f/10 or slower. If your telescope is faster than about f/8, you may need to use better auxiliary optics. I can definitely tell the difference between certain Barlow lenses when I use them on a 9.4 cm f/7 refractor, while the same lenses seem the same when used on a 9 cm f/11 or 20 cm f/10 Cassegrain.

Common defects introduced by auxiliary optics can include field curvature, color fringing, and spherical aberration. Severe coma and astigmatism are less common, except in the case of some stronger telecompressor lenses. (Aberrations are covered in chapter 4.)

The apparent performance of an auxiliary optic can also be influenced by aberrations in your telescope. If the optic has the same types of aberrations as your telescope (such as undercorrected spherical aberration), the results may not be very good; however, if the optic has aberrations which are opposite to those of your telescope (such as the optic being spherically over corrected and the telescope spherically undercorrected), using the optic may actually improve your results!

Following are some descriptions of specific auxiliary optics and equations for calculating your telescope's effective focal length when you use popular types of auxiliary optics. You don't necessarily have to calculate the exact answer to get good results; just being reasonably close (within a few percent) is good enough to facilitate a reasonably good exposure.

3.5.1 Telecompressors

Telecompressors (also called focal reducers) facilitate a shorter exposure time at the expense of a smaller image scale and possibly some vignetting. They are positive (convex in simple form) lens systems which are used between the telescope and focal surface. Telecompressors are usually specified in terms of relative magnification rather than focal length, but knowing the focal length of the telecompressor optics can improve the accuracy of your calculations. You can roughly determine the focal length of a telecompressor lens by measuring the distance from about the center of its lens group to the point at which it images distant objects when used by itself. Another method is to reverse engineer the system based on its specified reduction value and the distance from the focal surface at which it is normally used. The effective reduction of a given telecompressor is determined by its distance from the focal surface. Some telecompressors will perform well when used at varying distances from the focal surface, while others are optimized for a specific distance.

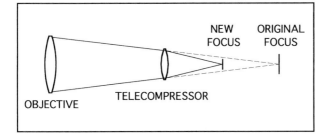

Figure 3.4. A telecompressor provides a shorter focal length and faster f/ratio.

There is a limit to how much reduction can be achieved with a given telecompressor system because the required back focus distance increases substantially when a telecompressor is used at a greater reduction, and this back focus distance can easily exceed the available back focal distance of many telescopes. In addition, too much reduction with a reflecting telescope will result in the telecompressor being moved closer to an infinity focus position where it may actually begin to image the telescope's central obstruction as a dark area in the center of the picture.

If extreme reductions are desired, a telecompressor system loosely resembling a fast f/ratio camera lens is typically the most effective. Accordingly, it usually works better to "stack" two achromatic telecompressors than to use a single relatively strong achromatic lens at an extreme reduction. Also, it is typically best for the weakest of the two telecompressors to be in front.

When telecompressors are stacked, it is usually a good idea for one of them to be a well corrected system such as a popular 0.63× reducer–corrector for a Schmidt–Cassegrain telescope. For the second lens, a focal length between 125 mm and 250 mm is usually best for use with film cameras, provided that the lens diameter is not more than about 0.35× its own focal length. Shorter focal length telecompressor lenses can be used with suitable video cameras and digital still cameras.

The illuminated circle gets smaller when more extreme reductions are used. In fact, the illuminated area can easily become substantially smaller than the film format area. Electronic imaging cameras can benefit from extreme reductions with telecompressors because the sensor size is usually quite small in relation to the image circle and popular film camera formats. In general, about 0.6× is the minimum practical reduction for the 35 mm film format, but reductions as low as 0.35× can be used if you can tolerate significant vignetting. For a smaller format electronic imaging camera, 0.25–0.35× is usually the minimum practical reduction.

I have prototyped CCD telecompressor systems having reductions down to 0.15×, but these (at least when made from off the shelf lenses) have had some problems with spherical aberration.

In general, a telecompressor will provide a reduction of 0.5× when it is used at a distance of half its own focal length from the focal surface. When so used, the required back focal distance will be as though the focal surface was twice as far from the telecompressor lens than the point at which the reduced image is actually formed. This is easy to envision when you realize that the light cone from the telescope aperture to an imaged point at the focal surface converges twice as fast behind the telecompressor lens when a 0.5× reduction is used.

For example, say you are using a telecompressor lens with a 130 mm focal length at a distance of 65 mm from the focal surface, and that it is being used behind an f/10 telescope. The f/10 (1:10 convergence ratio) light cone will change to f/5 as it passes through the lens. Therefore, the diameter of the light cone at the optical center of the lens will be 1/5 of the 65 mm distance from the lens to the focal surface; or 13 mm. In regard to the back focus setting of the telescope, the f/10 light cone would still apply, so the distance from the lens to the effective back focal point will be ten times the diameter of the light cone at the telecompressor lens; or 10 × 13 mm, or 130 mm. In other words, if the above telecompressor is inserted between the telescope and camera without changing the relative position of the camera, the focus will have to be changed by 65 mm. If you are using a telescope with a rack and pinion focuser, the focuser will have to be racked *inward* by this amount. If you are using an internally focusing telescope, you will have to adjust the focus so the image moves away from the back of the telescope by the same distance.

A reduction of 0.5× is very easy to envision and calculate in one's head. Other reductions are not as easy to envision. Accordingly, a less obvious situation of a 240 mm focal length lens used at a distance of 112 mm from the focal surface will be used to demonstrate the appropriate equation for determination of the reduction ratio:

$$1 - \left(\frac{\text{Distance to focal surface}}{\text{Telecompressor focal length}} \right) = \text{Magnification}$$

$$1 - \left(\frac{112 \text{ m}}{240 \text{ m}} \right) = 1 - 0.467 = 0.533$$

But wait, there's more! A second type of equation will also work to calculate the magnification:

$$\frac{\left(\dfrac{\text{Telecompressor}}{\text{focal length}} - \dfrac{\text{Distance to}}{\text{focal surface}}\right)}{\text{Telecompressor focal length}} = \text{Magnification}$$

$$\frac{(240\ \text{mm} - 112\ \text{mm})}{240\text{mm}} = \frac{128}{240} = 0.533$$

The focal length and numerical f/ratio are changed by the same factor as the magnification, so a conventional 2000 mm focal length f/10 telescope would work at a focal length of 1066 mm and an f/ratio of f/5.3.

The effective back focal distance behind the telecompressor can be calculated in a number of ways. You can either base the calculations on the diameter of the light cone at the lens, or use a more direct method as shown below:

$$\frac{\text{Distance to focal surface}}{\text{Magnification}} = \text{Back focal distance}$$

$$\frac{112\ \text{mm}}{0.533} = 210\ \text{mm}$$

This is 98 mm behind the actual focal surface (210 mm behind the telecompressor, which is 112 mm in front of the focal surface).

If the telescope is a popular internally focusing Cassegrain, the Cassegrain focal length will increase by about 3.8 times (the exact increase is dependent on a telescope's specific attributes so it can vary considerably) as much as the *additional* back focus required for the telecompressor, and this increase will apply to the final result. Accordingly, the effective focal length of the example telescope (before the effect of the telecompressor) will become about 2372 mm (2000 mm + [3.8 × 98 mm]) with an f/ratio of 11.9. After the 0.533× reduction by the telecompressor, the effective focal length is 1264 mm, with an f/ratio of 6.3. This situation is complicated further when the telecompressor cell increases the distance between the camera and telescope, since this additional back focus will also affect the telescope's focal length. From this example, one can see that a telecompressor designed to provide a reduction of 0.63× with a popular internally focusing SCT instrument will usually provide slightly more reduction with a telescope having a conventional external focuser.

3.5.2 Barlow Lenses

Barlow lenses provide a larger image scale at the expense of a slower f/ratio and longer exposure time. They are negative (concave) lens systems which are used between the telescope and focal surface. Like telecompressors, Barlow lenses are usually specified in terms of relative magnification rather than focal length. You can roughly determine the focal length of a Barlow lens by reverse engineering it based on its specified magnification and the distance from the focal surface at which it is normally used. The effective magnification of a given Barlow lens is determined by its focal length and distance from the focal surface.

Barlow lenses are a bit less picky about the acceptable range of distance to the focal surface than telecompressors, and back focal distance is not a significant concern. Also, Barlow lenses seldom reduce the illuminated field at the focal surface; in fact, they often increase the size of the illuminated circle. A strong Barlow lens can produce a higher magnification for a given back focal distance, and one strong lens typically works better than stacking two weaker ones; however, the amount of field curvature for a given magnification is typically more with a strong Barlow lens than it is for a weak one of the same basic design.

The limit to how much magnification can be achieved with a given Barlow lens is determined by the maximum distance that is practical to have between it and the focal surface. Many popular photographic teleconverters will work just as well as a Barlow lens for astrophotography with most telescopes. Some teleconverters may be subject to more flare and ghost images

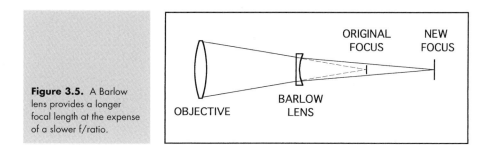

Figure 3.5. A Barlow lens provides a longer focal length at the expense of a slower f/ratio.

than a simpler Barlow lens, but a teleconverter typically has less field curvature and provides better off-axis images. High end teleconverters such as the Nikon TC-301 are excellent for many types of astrophotography. If a more extreme magnification is required than what you can reasonably get with a Barlow lens, it is usually best to resort to other arrangements such as eyepiece projection or afocal photography.

In general, a Barlow lens will provide a magnification of 2× when used at a distance of its own negative focal length from the focal surface. When so used, the required back focal distance will be as though the focal surface was twice as close to the Barlow lens than it actually is because the Barlow lens reduces the convergence angle of the light cone. This is easy to envision when you realize that the light cone will converge only half as much as that of the telescope when a 2× magnification is used.

For example, say you are using a Barlow lens with a –130 mm focal length at a distance of 130 mm from the focal surface, and that it is being used behind an f/10 telescope. An f/10 (1:10 convergence ratio) light cone will change to f/20 as it passes through the Barlow lens. Therefore, the diameter of the light cone at the optical center of the lens will be 1/20 of the 130 mm distance from the lens to the focal surface; or 6.5 mm. In regard to the back focus setting of the telescope, the case of an f/10 light cone would still apply, so the distance from the Barlow lens to the effective back focal point will be ten times the diameter of the light cone at the Barlow lens; or 10 × 6.5 mm, or 65 mm.

In other words, if the Barlow lens is inserted between the telescope and the camera without changing the position of the camera, the focus will have to be changed by 65 mm (assuming both the camera and Barlow lens are moved together by the focuser). If you are using a telescope with a rack and pinion focuser, the focuser will have to be racked *out* by this amount to achieve proper focus. If you are using an internally focusing telescope, you will have to adjust the focus so the image moves toward the back of the telescope by the same distance.

This is why back focus distance is seldom a problem with Barlow lenses; if the focuser won't move out far enough, you can simply add an extension tube. Most Barlow lenses and teleconverters include a tube of about the correct length to compensate for the focus change. In this case, the required focus adjustment is less than it would be if the Barlow lens had no extension tube.

A magnification of 2× is very easy to envision and calculate in one's head. Other magnifications are not as easy to envision. Accordingly, a less obvious situation of a –66 mm focal length lens used at a distance of 100 mm from the focal surface will be used to demonstrate the appropriate equation for determination of the magnification. For simplicity, positive numbers can be used in the calculation. The equation is like that used for a telecompressor, but the parenthetic elements are added rather than subtracted from the number 1:

$$1 + \left(\frac{\text{Distance to focal surface}}{\text{Barlow focal length}} \right) = \text{Magnification}$$

$$1 + \left(\frac{100 \text{ mm}}{66 \text{ mm}} \right) = 1 + 1.515 = 2.515$$

As with the telecompressor, a second type of equation is also useful for the Barlow lens:

$$\frac{\left(\begin{array}{c} \text{Distance to} \\ \text{focal surface} \end{array} - \begin{array}{c} \text{Barlow} \\ \text{focal length} \end{array} \right)}{\text{Barlow focal length}} = \text{Magnification}$$

$$\frac{(100 + 66)}{66} = \frac{166}{66} = 2.515$$

The focal length and numerical f/ratio are changed by the same factor as the magnification, so a 2000 mm focal length f/10 telescope would work at a focal length of 5030 mm and an f/ratio of 25. With an internally focusing Cassegrain telescope, the change in focal length due to modifying the back focus distance can change things a bit, but usually not as much as for a telecompressor.

The effective back focal distance behind the Barlow lens (i.e. where focus would be in relation to the Barlow lens position if the Barlow lens were removed without changing anything else) can be calculated in a number of ways. You can either base the calculations on the diameter of the light cone at the lens, or use a more direct method as shown below:

$$\frac{\text{Distance to focal surface}}{\text{Magnification}} = \text{Back focal distance}$$

$$\frac{100 \text{ mm}}{2.515} = 39.8 \text{ mm}$$

3.5.3 Eyepiece Projection

Eyepiece projection (sometimes called positive projection) can provide a radically larger image scale at the expense of a substantially slower f/ratio and longer exposure time. Here, an eyepiece is used as an enlarging lens for the telescope image. Two real images are formed; one at the focus of the telescope, and another at the focal surface in the camera. The image formed by the telescope alone is obviously located between the telescope and eyepiece optics. Eyepieces are typically specified in focal length, which can make calculating the photographic focal length easier. The effective magnification with a given projection eyepiece is determined by its distance from the focal surface.

Eyepieces are not too picky about the acceptable range of distance to the focal surface, and back focal distance is not a significant concern. Eyepiece projection seldom reduces the illuminated field at the focal surface, but some eyepieces can produce a small hot spot at the center of the picture when used with a reflecting telescope, particularly a Cassegrain. Such a hot spot is typically caused by unwanted reflections between the eyepiece and the secondary mirror of a Cassegrain telescope. If you get unwanted reflections in your pictures, it may be worth reversing the orientation of the eyepiece in your setup or using a different type of eyepiece.

The limit to how much magnification can be achieved with a given eyepiece lens is determined by the maximum distance that is practical to have between it and the focal surface. Magnifications well in excess of ten times a telescope's prime focal length are relatively easy to obtain with a short focal length eyepiece. It is usually best to use a good, short focal length eyepiece by itself rather than trying to use a longer one in combination with a Barlow lens.

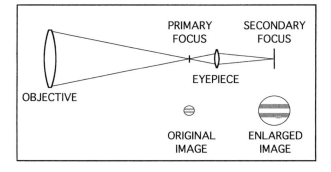

Figure 3.6. Eyepiece projection can provide a very long focal length.

It is very important that the eyepiece optics are clean when used for projection photography because diffraction from dust particles on its optics (or refraction through any oily film) can affect the image more significantly than it would in other types of photography.

Most eyepiece projection accessories include an extension tube which can minimize the amount of focus compensation that is required. In general, an eyepiece will provide a magnification of slightly less than 1× for every multiple of its focal length from the focal surface. This can provide a rough estimate of the magnification, but the equation below is more precise. The example will be for a 9 mm eyepiece which is used at a distance of 100 mm from the focal surface:

$$\frac{(\text{Projection distance} - \text{Eyepiece focal length})}{\text{Eyepiece focal length}} = \text{Magnification}$$

$$\frac{(100\,\text{mm} - 9\,\text{mm})}{9\,\text{mm}} = \frac{91\,\text{mm}}{9\,\text{mm}} = 10.1$$

The focal length and numerical f/ratio are changed by the same factor as the magnification, so a 2000 mm focal length f/10 telescope would work at a focal length of 20,200 mm (2000 mm × 10.1) and an f/ratio of 101 (10 × 10.1).

If the telescope is an internally focusing Cassegrain, the actual effective focal length may be substantially different than what calculations based on the telescope's specifications would indicate. For instance, if a conventional visual back and tele-extender tube are used on a 20 cm f/10 SCT, the Cassegrain focus will be only about 35 mm behind the rear cell of the telescope rather than at the standard distance of 105 mm. This new back focal distance will cause the telescope's effective focal length prior to projection to be shorter, resulting in a shorter projected focal length. Specifically, the 70 mm (105 mm – 35 mm) reduction in back focal distance will result in the Cassegrain focus being only about 1734 mm, which is 266 mm (70 mm × the back focus compensation factor of about 3.8) shorter than normal. Therefore, the effective projection focal length will be about 1734 × 10.1, or 17,513 mm.

"Eyepiece projection" can also be accomplished with optics other than telescope eyepieces. Alternatives include motion picture camera lenses, microscope eyepieces, and microscope objectives. Microscope objectives do not always work as well as one might expect because an aerial image (i.e. one not formed on a physical surface) is what is being projected. This type of image has different characteristics than what a microscope objective was designed for.

3.5.4 Projection Screen Images

Projection of a telescopic image onto a screen is occasionally used for solar observation. Here, the focal length and f/ratio are determined in the same manner as for eyepiece projection; however, calculating the exposure time is not always straightforward because the required exposure time can be influenced by factors such as ambient light and screen reflectivity. Accordingly, some trial and error may be in order for this type of photography, but you may get close to the right exposure by taking a reflective light meter reading of the screen projection and exposing for about one f/stop more than what the meter says.

I have had some success in using a good slide projector lens as a solar projection lens. The relatively long focal length of such a projector lens allows the image from a telescope to be projected some distance (up to several meters) into a well shaded area. This in turn allows the image to be projected into a house through a carefully guarded window or doorway. In addition, a good slide or movie projector lens can be more tolerant of heat from the solar image than most eyepieces. Screen projection can facilitate solar photography with surprisingly modest optics. Even some binoculars can be used for solar projection if you are careful, put an opaque cap on the side you are not using, don't try to look through them, and don't leave them pointed at the sun for more than a few seconds at a time. Masking all but the central third or so of a binocular objective can reduce the risk of internal damage to the binoculars from the concentrated sunlight.

In any form of solar imaging, it is important to be sure to position your telescope, binoculars, or other optics so people cannot inadvertently look directly into the eyepiece when the objective is inadequately filtered. Screen projection requires particular caution because you have to ensure that people don't stick their head between the eyepiece and screen in an effort to look directly through your telescope or binocular.

Projection screens are often utilized in public astronomy events, and such events can provide a fun way to share your hobby with others. Unfortunately, there are times when a few unruly members of the public (or their undisciplined children) may make things difficult for everyone concerned. These days, some people seem to act in a careless and presumptuous manner, and a few may even try to come up to your equipment and start handling it as though they owned it. A rare specimen may even try to walk away with an eyepiece, so the owner of a telescope may need to be increasingly vigilant at public events. Things sure aren't like they used to be.

3.5.5 Afocal Imaging

Afocal photography is one of the most straightforward types of imaging that involves auxiliary optics; just point your camera and lens into the telescope eyepiece. It has the advantage of being compatible with a variety of cameras and telescopes, including some cameras having fixed lenses and some telescopes having fixed eyepieces. Like eyepiece projection, afocal imaging can provide a radical increase in image scale at the expense of a substantially slower f/ratio and longer exposure time. Some vignetting and field curvature is likely with afocal imaging, but this usually is not a problem for photography of planets and other narrow angle objects.

In afocal imaging, the camera lens aperture should be wide open. If the lens is stopped down, excessive

Figure 3.7. This camera and lens are squarely positioned near a telescope eyepiece for afocal photography. Tilting the eyepiece toward the side as shown permits it to move directly away from the camera as the telescope tracks.

vignetting may occur. It is also important for the camera lens to be relatively close to the eyepiece and centered directly behind it. Ideally, the exit pupil of the eyepiece will be imaged near the plane of the camera lens aperture blades, but this optimum situation is not always possible due to the close proximity of the exit pupil to a typical eyepiece.

For afocal imaging, it is not necessary for the camera to be attached to the telescope. In fact, mounting the camera on a separate tripod can reduce vibration of the telescope and eliminate the need for SLR mirror lockup or special counterbalancing arrangements. It is helpful to orient the telescope so the eyepiece moves directly toward or away from the camera as the subject is tracked. This can keep you from having to reposition the camera as often, but if the telescope is oriented so that its eyepiece moves toward the camera, it is important to be sure you don't let the sidereal drive run unattended for so long that the eyepiece touches the camera lens.

If you do not have a sidereal drive, you can position the center of your subject at the edge of your eyepiece field which is opposite to the direction of drift, then wait for the subject to drift near the center before you take your pictures. It may be possible to minimize drift during your exposure by gently flexing the tripod or lightly pushing or pulling on the eyepiece. You can practice these methods and preview their effectiveness by looking at the subject in your SLR camera viewfinder.

It is helpful to orient your camera tripod so the camera can be repeatedly moved to and from the position near your eyepiece when you reposition or focus the telescope. This can be effectively accomplished by setting up so you can swing the camera away from the telescope by tilting the tripod so that only one leg leaves the ground. Another tip is to use a dark cloth or a black cardboard cone or disk around the eyepiece in order to reduce interference from nearby light sources.

The change in effective focal length which results when a given eyepiece is used for afocal imaging is determined by the focal length of the camera lens, and is directly proportional to the ratio between the focal length of the camera lens and eyepiece. For instance, if a 50 mm camera lens is used to take a photograph through a 9 mm eyepiece, the magnification is 50/9, or 5.55×. To get the final focal length and numerical f/ratio, simply multiply the telescope focal length and f/ratio by this magnification.

Another way to calculate focal length when using the afocal method is to just multiply the focal length of your camera lens by the visual magnification of the telescope. For instance, a 2000 mm focal length telescope with a 9 mm eyepiece produce a magnification of: 2000/9 = 222×. If you use a 50 mm camera lens with this combination, the focal length will be 222 × 50 mm = 11,100 mm. To get your f/ratio, just divide this focal length by the telescope aperture. For example, a 20 cm aperture telescope would work at f/55.5 (11,100 mm focal length / 200 mm aperture).

If you are using a Barlow lens between the telescope and eyepiece, its magnification will apply to your final result, so if you use a 2× Barlow lens with the above setup, the telescope's original image would be magnified by about 11 times (2 × 5.55), providing a focal length of 22,200 mm.

Some interesting interrelations can be applied as shortcuts in your calculations. For example: The difference in the f/ratio is directly proportional to the difference in image size, so it is also proportional to the difference between the focal length of the camera lens and eyepiece. The final f/ratio can also be calculated directly by simply multiplying the original telescope f/number by the magnification of the telescope's original focal length.

Another way to calculate the f/ratio is to divide the camera lens focal length by the diameter of the eyepiece exit pupil. For example, the exit pupil of a 9 mm eyepiece when used on an f/10 telescope is 0.9 mm (9 mm / 10). With a 50 mm camera lens, the final f/ratio would be 50/0.9 = f/55.5. This can be a time saver, since the f/ratio is all you need to know at the time you take the picture (the f/ratio is what determines the required exposure time for a given subject and film). The focal length can always be calculated after the fact.

The limit to how much magnification can be achieved with a given eyepiece or lens is determined by the maximum practical ratio between the eyepiece and camera lens focal lengths. Magnifications in excess of ten times the telescope's prime focal length are relatively easy to obtain with a short focal length eyepiece and a normal or telephoto camera lens.

It is very important that the eyepiece and camera lens optics are clean when they are used for projection photography. Diffraction from dust particles on the optics can affect the image more significantly than they would in other types of photos. Larger dust particles can even cast visible shadows on your image.

Focusing for an afocal image can be relatively easy and foolproof. One method is to use a binocular or monocular of about 7–10 power as a focusing magnifier, and focus by looking through the telescope eyepiece with the binocular rather than trying to focus through the relatively dim camera viewfinder. The magnification of the binocular or monocular is required to compensate for the accommodating effects of your eye when you look into an eyepiece. First, look directly at the stars through your binoculars and adjust the focus until the image looks sharp. Then, locate and center your subject (or star near it) in your telescope. After this, orient your binocular so that you can look into the telescope eyepiece through it, and while looking, adjust the focus of the telescope (not the binocular) until the image looks sharpest. This will focus the telescope for any camera and lens which is focused on infinity!

If your telescope has been outside for less than half an hour or if the temperature is changing rapidly, you may want to double check the focus every few minutes. The focus of some telescopes can shift as they cool.

Finally, it is worth pointing out that afocal imaging can also be used to reduce the telescope's effective focal length if the focal length of your eyepiece is longer than that of your camera lens. Such reductions are not practical for most film photography because vignetting usually results when a wide angle lens is used on the camera; however, the afocal method can provide a very practical way to get a wider field and faster f/ratio with a video camera or other electronic imager. For instance, if you use a short 10 mm focal length video lens with a long 50 mm focal length telescope eyepiece, you can get an image scale that is only 0.2× the size you would get by coupling the camera body directly to the telescope! This provides similar results to a telecompressor, but the afocal method is admittedly more subject to flare and unwanted reflections than a good telecompressor.

Afocal imaging requires a minimum of equipment, but the results can be spectacular. You can use it with film cameras (even some fixed lens cameras), video cameras, intensified cameras, and integrating electronic cameras. It even works with most zoom lenses!

3.6 Light Transmission Value

The transmission value of a telescope is what really matters when calculating your exposure time. As light makes its way through a telescope, some of it is obstructed, reflected, scattered, and absorbed. You don't need to know the exact transmission value of a telescope to get good astrophotos, but it is good to at least have a general idea.

A refracting telescope will typically transmit between 82 and 96 percent of the available light; a Newtonian reflector between 71 and 90 percent (depending on the coatings and obstruction); a typical Cassegrain between 66 and 87 percent, and a typical commercial Cassegrain having a corrector lens between about 61 and 83 percent.

Variations for refractors relate mostly to lens thickness and any antireflection coatings which may be used on optical surfaces which are exposed to air. Variations for reflectors relate mostly to obstruction size, efficiency of reflective coatings, whether or not corrector lenses are used, and whether or not the corrector lenses have antireflection coatings.

The transmission values mentioned above are all the information you may need for most astrophotography, but some may want to calculate the transmission value of their own particular telescope. If you do, the following analysis will allow you to crunch numbers until your heart is content.

The examples are for a telescope without auxiliary optics such as Barlow lenses and the like. Such auxiliary optics have relatively good transmission characteristics. Therefore, you can just use the transmission value for your telescope as a baseline when using most auxiliary optics. Diagonal mirror attachments usually have losses of well under 10 percent due to the comparatively greater efficiency of reflective coatings at a 45 degree incident angle (as opposed to their efficiency at the normal angle of incidence encountered in a primary mirror, for example). Diagonal prisms having antireflection coatings on the entrance faces are also relatively efficient because the internal reflection of a diagonal prism is total, provided there is no coating on the hypotenuse.

Some telescopes have obstructions in the aperture. The most common source of obstruction is a secondary mirror and related baffling. Other telescopes (such as refractors) are unobstructed, so they do not have losses from an obstruction; however, a refractor will have losses from other factors such as reflection and absorption. With some glass, absorption can be as much as one percent per centimeter of thickness.

A significant amount of light can be lost by reflection from a refracting surface which is exposed to air. Such a surface is typically called an "air to glass" surface. The loss due to reflection is typically about 4 percent per surface for an uncoated lens, 1.5 percent per surface if magnesium fluoride antireflection coatings are used (these usually have a blue or purple reflection), and 0.5–1.5 percent per surface for multicoatings (which often have a red, green, or amber reflection). The actual loss by reflection is dependent on wavelength, so these figures are just provided for reference.

Reflective mirror coatings can scatter and absorb light. The amount of loss for a given coating depends on wavelength and the angle of incidence. In general, the loss is greatest when you are looking straight into a mirror. Therefore, a diagonal mirror will tend to provide a brighter reflection than a primary mirror with the same coating.

A primary mirror with ordinary but new aluminum coatings will reflect about 89 percent of the light reaching it, resulting in an 11 percent loss. Enhanced aluminum can increase the reflection to 96 percent or more, and a good enhanced silver coating can reflect more than 98 percent.

The efficiency of most reflective coatings is dependent on wavelength, but most popular telescope mirror coatings are optimized for good performance throughout the visual range. Some enhanced coatings also have relatively little scattering, which can result in higher contrast. Rhodium is a durable coating with even less scattering than some enhanced coatings; however, it is seldom used in modern amateur instruments because of its typical reflectivity of 85 percent or less.

Now that the major sources of light loss have been addressed, we can run through a few examples. It may be enlightening to compare the difference between a 20 cm aperture Schmidt–Cassegrain telescope having ordinary coatings to one with enhanced mirror coatings and magnesium fluoride coated corrector plate.

All of the losses are cumulative, so the best accuracy will be obtained by calculating the loss at each surface based on the light reaching it. Only a rough figure can be obtained by just summing all of the losses. You can use any units you like in the calculations, so long as you are consistent. I often use millimeters just because I'm used to it, but the following example will use larger units.

A typical 20 centimeter aperture f/10 Schmidt–Cassegrain telescope has a secondary obstruction diameter of about 7 cm; about 35 percent of the aperture diameter. The light loss from the obstruction can be calculated by comparing its area to that of the telescope aperture. Since we are only after a relative figure, there is no need to calculate the actual surface area via using pi multiplied by the square of the radius. Therefore, we can just square each respective number (pretending for a moment that the aperture is square) to get their relative areas: 20×20 cm aperture = 400 square cm; 7×7 cm = 49 square cm.

To get the effective area, just subtract the obstruction area from the aperture: 400 – 49 = 351 square cm.

To get the decimal value of the obstruction, divide the area of the obstruction by total area: 49/400 = 0.122, or a loss of 12.2 percent.

To get the effective aperture (i.e. the diameter an unobstructed system having an equivalent aperture area would have), take the square root of the effective area: Square root 351 = 18.7 cm equivalent effective aperture; versus the original 20.0 cm.

To get the transmission value, divide the focal length by the effective aperture: 200 cm focal length / 18.7 cm effective aperture = f/10.7; versus the original f/ratio of f/10.0.

Not bad so far, but we still have to compensate for losses in the optics. In the case of a telescope with no antireflection coatings on the corrector, each surface of the corrector will cause a loss of 4 percent. To get the transmission of the corrector surface, the 4 percent loss is subtracted: 100 – 4 percent of 100 percent = 96 percent. Therefore, 96 percent of the light gets past the first surface.

Of the 96 percent of the light that remains, only 96 percent will get through the second surface. This can be calculated in a number of ways, depending on whether your point of reference is the percentage of transmission or the percentage of loss. The following example is referenced to the percentage of transmission:

0.96 (decimal value of transmission) multiplied by 0.96 = 92.2.

Still not bad, but we still have to deal with the mirror coatings. With conventional mirror coatings, the loss can be substantial. We start with the 92.2 percent of the light we have left:

> 0.922 multiplied by the 0.89 reflectivity of the primary mirror = 0.821.

Then:

> 0.821 multiplied by the 0.89 reflectivity of the secondary mirror = 0.731.

As you can see, the optics of our Schmidt–Cassegrain telescope with conventional coatings only transmit about 73.1 percent of the light! When we throw in the 12.2 percent loss from the obstruction, only 64.2 percent of the light is left! Note the answer is substantially different than it would be if we simply subtracted all of the losses as percentages from 100 percent: 100 – (4 + 4 + 11 + 11 + 12.2) = 57.8 percent. A transmission of only 64 percent may sound bad, but it can be compensated for by increasing the exposure time to about 1.5 times of what would be used for a theoretical "loss-less" system.

To get the new equivalent aperture, square the actual aperture (20 × 20 cm = 400) and multiply the answer by the decimal value of the transmission (400 × 0.642 = 257), then take the square root of the new answer: square root 257 = 16.0 cm effective aperture. From this exercise, we can see that, in terms of light grasp, our 20 cm Schmidt–Cassegrain telescope would offer little advantage over a 15 cm refractor!

To get an f/number equivalent of the transmission value, divide the focal length by the effective aperture: 200/16 = f/12.5. The true f/ratio is still f/10, but the amount of light reaching the focal surface is equivalent to that of a theoretical 100 percent efficient 16 cm f/12.5 telescope. This "transmission value" is typically indicated as "t/12.5" or "t = 12.5".

Now, let's see what happens if the Schmidt–Cassegrain telescope has enhanced coatings. We start out with a 12.2 percent loss from the secondary obstruction, just like in the previous case.

The transmission of the coated corrector is 98.5 percent per surface, so 0.985 × 0.985 = 0.970. The reflectivity of the enhanced aluminum mirrors is at least 96 percent, so 0.970 × 0.960 = 093.1 and 0.931 × 0.96 = 0.894. Therefore, the optics transmit 89.4 percent of the light; more throughput than the primary mirror surface alone in the previous example! To get the transmission including effects of the secondary obstruction: $0.894 \times 0.878 = 0.785$. And to get the effective aperture: Square root $(400 \times 0.785) = 17.7$ cm; over 10 percent more effective aperture than the same telescope with standard coatings! The telescope with enhanced coatings has an effective transmission value of: $200/17.7 = t/11.3$.

If you want to know the difference in required exposure value with both telescopes, the exposure for the one having conventional coatings would have to be: 78.5 percent transmission of the enhanced system divided by the 62.4 percent transmission of the standard system = 1.26. This means that an exposure through the telescope with standard coatings would have to be 1.26 times longer than that required for the same telescope with enhanced coatings, excluding the effects of reciprocity failure in the film.

This data has been provided in order to allow you to calculate exposure values for your astrophotography. You don't have to come up with exact specifications for your telescope in order to take good astrophotos, but knowing the basics of these calculations and recording your exposure data can help you more easily repeat equipment setups which provide the best results.

A sample form for recording exposure data is located in the appendix of this book. Using it as a template for consistently keeping track of your exposure data (in conjunction with the astrophotography exposure guide which is also in the appendix of this book) should assist you in accumulating enough information to photograph your favorite subjects without a lot of trial and error.

Chapter 4
What to Expect from Your Equipment

Most types of astrophotography can be mastered through trial and error or by learning a few specific techniques; however, learning the basics of how optics work can allow you to improvise and anticipate what level of results you can expect under various conditions. Learning such basics without resorting to trial and error can also save you a lot of time and film.

High optical quality is always desirable in principle, but surprisingly, really good optical quality is not absolutely essential for every type of astrophotography. Larger and sharper telescopes of a given design tend to cost more, though there are a few notable exceptions to this relationship. Therefore, it is a good idea to determine the minimum image quality that will be acceptable for your applications, particularly if finances are an issue or if astrophotography is only a casual hobby. There may be little point in investing thousands of dollars in a big and expensive telescope if you only want to use it a few times a year to shoot pictures of the moon. If you live in the city and plan to use your telescope at home most of the time, you may want to consider a smaller instrument, since a small, easy to use telescope tends to get used more often than a large or cumbersome one.

There are some types of astrophotography in which less than snappy image quality can and must be tolerated in spite of the quality of optics used. This is particularly true for a subject having such a small angular size that its image fills only part of the format even when an extremely long focal length is utilized. Here, the original image must be enlarged significantly in order to get a decent printed image size. This in turn affects the sharpness of your final printed image.

A radical increase in focal length may at first blush seem like it will solve the problem of imaging a subject having a small angular size, but there are many factors which limit how much the focal length should be increased. These factors include vibration (induced by wind or the camera shutter), tracking error (from imperfections in the sidereal drive), bad "seeing" (from atmospheric turbulence), optical performance (which has its limits), and the imposition of a slower f/ratio (which will increase the required exposure time). These issues are addressed below.

4.1 Vibration and Tracking Error

Vibration can spoil photos taken with even the best optics. A long focal length telescope has a relatively narrow field of view and provides a large image scale. This magnifies any motion or vibration of the optical system, so a stable mounting is very important. Even with a stable mounting, visible vibration may result from the camera shutter, wind, passing vehicles, and in some cases, even footsteps on the nearby ground.

To overcome vibration from a camera shutter, you can use an adequately sized "black card" as a shutter in front of the telescope. First, hold the card just in front of the telescope so it covers the entire aperture, then open the camera shutter with a cable release (or use the "T" setting if your camera has it). When vibration from the shutter settles down, move the black card (without touching the telescope) to uncover the aperture, then replace it at the end of the exposure. After this, close the camera shutter. There is a limit to how fast a "shutter speed" can be implemented with the black card technique, but it is certainly useful for exposures lasting half a second or more. With the "black card" technique, you don't need to lock up an SLR camera's mirror.

Many astronomical objects require a long exposure time and must be tracked (usually with an equatorial mount) as the film is exposed. Errors in tracking will cause the image to move on the film during the exposure, resulting in a streaked image. Equatorial mounts are covered in Chapters 2 and 7.

4.2 Atmospheric Conditions

Turbulence and other atmospheric effects can degrade astronomical images. While air does not refract light as much as a typical lens, it does refract it enough to influence telescopic images. Air masses of differing temperature and humidity have different densities, and as a result, differing refractive indices. The problem stems from the fact that the air is not always still, so air of differing temperatures is constantly moving and mixing; furthermore, it is doing all this between your telescope and subject! Such "seeing conditions" can cause the image in a telescope to shimmer or look blurred. You may even be able to detect wind direction a kilometer or so above your site by just looking at a planetary image in your telescope, since some seeing conditions can cause a planetary image to be distorted in a way that is reminiscent of looking through moving water to see a rock on the bottom of a stream. Atmospheric seeing conditions are also what causes stars to scintillate or twinkle. Planets usually do not twinkle because their angular size exceeds the changing angles at which localized turbulence can bend the light.

Not all turbulence is caused by weather systems or geological features. Substantial turbulence can result from smaller objects such as the roof of a house, convection above a busy street, heat from people standing near the front of your telescope, or even the disturbance of air as it passes by an observatory dome slit.

If you live in an area subject to excessive turbulence, there is often a lot to be gained by selecting a significantly higher observation site. Many of the best observatories are located in mountainous areas which have a relatively dry climate or are close enough to a western continental coast to take advantage of the relatively laminar air flow which can blow in from the ocean.

When selecting an observing site, it is important to consider the effects of localized turbulence. Some mountain ranges such as the Rocky Mountains can produce so much turbulence that one can often experience better seeing by taking the opposite approach and actually going to a lower elevation site which is down on the plains, several dozen kilometers east of the mountains.

Even when the air is still, atmospheric refraction can affect the apparent location of a celestial object which is near the horizon, making it appear to have a slightly higher elevation angle than it would have if there were no atmosphere. The effect can be so extreme near a low horizon that light from the bottom of the rising moon is refracted substantially more than light from the top, which in turn makes the moon look flattened as it rises or sets. When an object is this near the horizon, the earth's atmosphere also absorbs and scatters an increasing amount of its light, making it appear dimmer. The effect is most pronounced for shorter wavelengths such as blue light, so it is not uncommon for the rising or setting moon to appear orange or yellow. Another effect of atmospheric refraction includes color fringing in telescopic views of subjects having a low elevation angle, caused by a slight prismatic effect. Just like a simple lens, the atmosphere refracts shorter wavelengths of light at a greater angle than longer wavelengths.

Moisture in the air can affect observing as well. Clouds and fog are common problems, but even on a clear night, moisture can condense on your telescope optics. Deep sky photos are particularly sensitive to the effects of dew on your optics because even a little can cause excessive blooming around star images. To prevent the formation of dew on your optics and shield them from stray light, your optics should be shielded by a front tube having a length at least as long as the aperture of your telescope. Adding a weak heating wire near the front perimeter of your optics can also be helpful. The inside of the shield should be black in order to cut down on internal reflections. A simple Newtonian telescope having a closed tube usually does not require an added dew or light shield (also called a lens hood or dew cap), but a telescope having a front lens can be another matter. Refracting and Cassegrain telescopes are the most likely to require a dew shield, and some even include one as standard equipment.

A simple dew shield can be made from flexible cardboard or plastic and attached to your telescope with an elastic cord, Velcro, or other means. If you make your dew shield as a flat piece of material which can be wrapped around your telescope and add Velcro to the ends (where the material overlaps when wrapped around your telescope) it will be more portable because it can be stored flat or rolled up. You can even crease a flexible dew shield so it will fold small and flat enough to be stored in a case with your telescope or star charts, but this may cause the shield to be out of round unless you add a rigid clip over the front end of each crease when it is installed on your telescope. I have made and used this type of dew shield since the mid 1980s and they work pretty well. Some commercial vendors have since offered similar items, but these cost many times more than it would cost to build your own dew shield.

If your telescope is relatively small, you can use a piece of plastic pipe or something like a mailing tube cap or coffee can for a dew shield. Such a shield won't be as portable, but it can be rigid enough to accept some sort of counterweight for your camera; or maybe even serve as a counterweight itself. Counterweights are often necessary with small telescopes in particular. The telescope and piggybacked lens shown near the end of Chapter 2 each have a relatively simple home made dew shield attached. The larger one is heavy enough to nearly counterbalance a rear mounted camera, so the only other counterweight I needed was the small home-made one (which is adjustable in three axes) under the telescope to counterbalance the piggybacked lens on top.

Atmospheric effects often limit the resolution and efficiency of ground based telescopes, and most of us have to settle for using our telescopes on or near the ground. Airborne astronomy is occasionally utilized in the quest for high altitude, but for amateur astronomers, the usual motivation is simply to avoid clouds. Unfortunately, the relatively poor quality of aircraft windows keeps one from enjoying any benefits the higher altitude could offer in terms of seeing conditions. Notable times that amateur astronomers take to the sky include expeditions to get above the clouds and observe a total solar eclipse.

Airborne astronomy with a "real" telescope is utilized by only a few professional astronomers who have access to specialized platforms such as the now retired Kuiper Airborne Observatory or the proposed larger SOFIA, POST, and ALAT telescopes. Other ways of eliminating atmospheric seeing problems include space telescopes. These are expensive, but the Hubble space telescope has shown that they can also be effective.

4.3 Aperture, Optical Quality, and Resolution

There is no such thing as an optical system that will image a point light source of infinitely small angular size as an infinitely small point at the focal surface, so one may say that there is no such thing as a "perfect" lens or telescope. Even if a telescope could be made with perfectly smooth optics of exactly the right figure, a number of factors would still contribute to degrading the image. These include diffraction effects from the aperture, secondary obstruction (if any), and dust particles; scattering from optical surfaces and coatings; and atmospheric effects.

For the most part, all of these effects except atmospheric conditions tend to favor the use of a large aperture telescope. Unfortunately, atmospheric conditions can often keep you from realizing the higher resolution of a well made large aperture telescope; but on those nights that the atmosphere cooperates, such a large telescope can provide truly impressive high resolution images.

4.3.1 Aperture and the Theoretical Limit of Resolution

Angular resolution of a well made telescope is determined by aperture size, and the relationship is linear; that is, if you increase the diameter of the telescope aperture by a factor of two, you will increase the angular resolution by a factor of two. This is easy to envision when you consider that the absolute size of a telescope's Airy disk (the smallest spot of light that can be formed by an optical system) is dependent on f/ratio, not focal length.

The absolute size of the Airy disk is smaller for a fast f/ratio system than it is for a slow f/ratio system. Therefore, given the same f/ratio, the Airy disk will have the same actual size regardless of the aperture. A larger aperture telescope of a given f/ratio has a longer focal length and a larger image scale, so the Airy disk formed by the larger aperture telescope will have a smaller angular size and thereby be smaller in relation to features in the imaged subject. Therefore, more *angular* detail can be imaged by larger aperture telescope.

Since the maximum resolution of an optical system is limited by the above factors, it is not necessary to strive for absolute perfection in the optical figure. The accuracy you need will depend on your application.

If you will just be using the telescope at low magnification to observe and photograph the moon and deep

Figure 4.1. Left: Under good seeing conditions, a star image looks like this simulated Airy disk and diffraction ring pattern when viewed at high magnification through a quality optical system. The Airy disk is very small at the prime focus of a telescope; usually smaller than the film grain. Poorly made telescopes may produce a blob of light which is substaintily larger than the Airy disk, resulting in lower resolution. Center: Plot of Gaussian distribution of the Airy disk, with surrounding diffraction rings. Right : Illustration of the Rayleigh limit. Here, the Airy disks from two stars of equal brightness overlap slightly, but they can still be observed as two separate points. The angular size of the Airy disk is determined by the aperture size of an optical system. The physical diameter of the disk is determined by f/ratio. The Airy disk from an amateur telescope is many times larger than the angular size of a star, so such a telescope cannot reveal a star's true angular size.

sky objects, you may be able to get by with mediorce optics. If you want to emphasize observation and photography of planets and double stars, you will need a particularly well made telescope which is capable of resolving to what is called the "theoretical limit". The term "theoretical limit" is typically considered to be the Rayleigh limit in this section, though there are occasions when the same term may be used to describe a situation in which all undiffracted light from the telescope objective just barely falls within the Airy disk.

The Raleigh limit represents a unit of resolution somewhat smaller than the actual diameter of the light spot (or Airy disk) produced by a telescope. It is often said that achieving the Rayleigh limit requires a wave front accuracy of 1/4 wave at the focal surface. For reference, the theoretical limit of resolution at a wavelength of 550 nanometers (greenish yellow light) is about 1.2 arc seconds for a 10 cm telescope and 0.6 arc seconds for a 20 cm telescope. An arc second is 1/3,600 of a degree.

On a graph showing the light intensity of an in focus star image in the vertical dimension and distance perpendicular to the optical axis in the horizontal dimension, a representation of the Airy disk would have an outline similar to the shape of the top part of bell. Outside the Airy disk are a series of concentric and progressively dimmer diffraction rings.

The Airy disk is brightest in the center, so it is possible to resolve finer detail than the diameter of the entire disk. The Rayleigh limit is the separation required between two identical Airy disks for the light intensity at their intersection to be about 74 percent of that at their peaks. This arbitrary but easily detectable visual threshold is commonly used in amateur circles for evaluating optical performance. A raw photographic image may have substantially less resolution

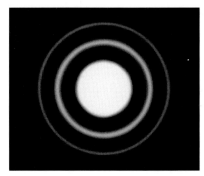
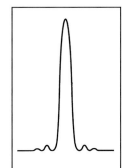
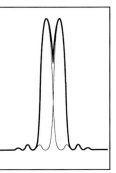

than what one can observe visually, depending on the film, image scale, and other factors.

4.3.2 Optical Quality and Resolution

The *wave front error* must be about $\frac{1}{4}$ wavelength or less if enough light is to fall within the Airy disk to enable the telescope to achieve the theoretical limit. To obtain a resolution as good as the theoretical limit, the optical surface accuracy must be at least as good as about $\frac{1}{2}$ wave for a refractor and $\frac{1}{8}$ wave for a reflector. The figure for the reflector must be more accurate because the wave front error is twice that of the mirror surface error by virtue of the fact the light is reflected. Less accuracy is required for a refractor because each optical surface refracts (bends) the light to a lesser degree than the local angle of the optical surface at a given zone.

This is not to say that you will get the best possible results by using telescopes that barely have this degree of accuracy; the optical figure can distort at different temperatures and the overall effects of optical surface accuracy and atmospheric distortion are cumulative. Therefore, it is not uncommon to be able to see the difference between optics producing a wave front accuracy of $\frac{1}{4}$ wave and those providing a wave front error of 1/10 wave or less. Even so, there is a limit to how good optics need to be for most applications. For instance, you may not want to go out and mortgage your house so you can get a telescope with 1/40 wave optics if you can more easily afford one of the same size with 1/10 or 1/20 wave optics.

If you will only be photographing objects at the prime or Cassegrain focus of your telescope (i.e. if you won't be enlarging its original image with a Barlow or other optic), you can get away with much less optical accuracy. It is important to realize that even a "bad" telescope can produce good photographs at prime focus; the optics only become a significant limiting factor when the image is magnified to the point that the spot size (whether it is an Airy disk or a blob of confusion) begins to exceed the size of the film grain.

It is amazing how bad an optical system can be and still meet this criterion. For example, my brother and I have used a 6 cm f/11.7 refractor having a maximum resolution about twice as bad as the theoretical limit (meaning it may have an optical figure as bad as a full wavelength), yet it can produce images that are grain limited on most color films. This is not surprising when you consider that very few camera lenses can resolve the theoretical limit either (particularly when used at full aperture), yet pictures taken with them still look sharp.

The effects of the film resolution and a telescope's optical resolution are cumulative, but not strictly additive; that is, a film resolution of 1/40 mm and an optical resolution of 1/40 mm will not usually produce a combined resolution as bad as 1/20 mm. This is partly because many optical defects cause only a small percentage of additional light to fall outside the Airy disk. Of course, photos taken with mediocre optics will show visible defects if a Barlow lens or teleconverter is used to increase the effective focal length. This is because a Barlow or teleconverter will magnify both the image of your subject and the optical defects of your telescope.

4.3.3 Diffraction from Obstructions

Diffraction is another consideration. About 84 percent of the light reaching the focal surface of a good unobstructed telescope is within the Airy disk, but a telescope of the same aperture which has a central obstruction will produce a slightly dimmer Airy disk. This is due in part to light loss from the obstruction, but some light reaching the film is also spread out by diffraction from the boundary of the obstruction. This diffracted light will be distributed throughout the surrounding diffraction rings, making some or all of them brighter. Enlarging the relative obstruction will tend to produce a still brighter diffraction ring pattern and an even dimmer Airy disk. The Airy disk formed by an obstructed system is actually a little smaller than that from an unobstructed system of the same f/ratio, but this does not result in a better planetary view because the brighter diffraction rings of the obstructed system tend to degrade the image contrast.

At prime focus, the inner (and brighter) diffraction rings usually cause the overall star image to become enlarged in a photograph because most film will tend to record the tiny image of the Airy disk and the inner diffraction ring(s) in a bright star image as a single spot. The size of the diffraction ring pattern corresponds to the size of the Airy disk, so the apparent disk enlargement may be negligible for wide field photos taken with fast f/ratio systems. (A fast f/ratio system produces a smaller Airy disk and diffraction ring pattern than a slow system.) In fact, halation will typically be a bigger problem than diffraction rings with a fast f/ratio system. Therefore, a fast astrograph can have an obstruction diameter almost half as large as its aperture and still produce good results.

The effects of a large obstruction are more obvious when a slow f/ratio is used for planetary or other imaging. Here, any extra light in the diffraction rings will tend to visibly lower contrast because the diffraction pattern is larger in relation to both the imaged subject and the film grain. An obstruction diameter of up to 15 or 20 percent of the aperture will not have too serious an effect on planetary images, but one over about 33 percent can significantly affect performance at a long effective focal length.

A spider such as that used to support the diagonal mirror of a Newtonian telescope also introduces diffraction. Each diffraction spike is oriented perpendicular to the boundary of the vane causing it, so a spider having two straight vanes which are oriented at a right angle to each other will produce two intersecting diffraction lines (or a total of four spikes) having the same perpendicular orientation to each other. Since the spikes are from diffraction, they will appear on both sides of the star image even if a spider vane is on only one side of the central obstruction. For example, a three vane spider in which the vanes are 120 degrees apart will produce a star image having three intersecting diffraction lines, for a total of six spikes.

Diffraction spikes tend to be long and thin if the spider vanes are thin, and short and bright if the vanes are thick. In general, a thin vane is best for planetary photography and other high resolution imaging because a smaller portion of the long diffraction spike from a thin vane will overlap a planetary image, causing less reduction in contrast. Many serious astrophotographers tend to go for a maximum spider vane-width of about one percent of the aperture diameter rather than striving for potentially problematic paper thin vanes.

The boundary of an obstruction or aperture is what causes diffraction, so the structure of a diffraction spike is similar to what the radial cross-section of diffraction rings would be for a star image formed by an aperture having the same diameter as the width of the spider vane obstruction.

Some amateur telescope makers have sought to eliminate diffraction spikes by using curved spider vanes. One trick I tried some time ago was attaching masks which have serpentine or scalloped boundaries to a straight spider. Such efforts are only necessary if you are trying to eliminate spikes in your pictures. In practice, diffraction spikes are only obvious in long exposures which include relatively bright stars. They seldom pose a problem for planetary photography.

4.3.4 Optical Aberrations

Inaccuracy in an optical surface may encompass a variety of forms. Each form of inaccuracy will tend to produce a particular type of image defect.

Spherical aberration is one of the most common aberrations. It is caused by the inner radial zones of an optical system bringing light to focus at a different point than light from the outer zones. Spherical aberration can take the fun out of planetary observing because it produces an image which sort of mushes in and out of focus. A star image may look round, but it never really forms an good Airy disk. Most telescopes with moderate spherical aberration will provide acceptable photos of most objects at prime focus if the image defects are small, but telescopes with extreme aberrations are a different story.

Eliminating spherical aberration is why a Newtonian telescope has a paraboloidal primary mirror instead of a spherical one. A spherical mirror produces a perfect image of something at a distance as close as its radius of curvature (twice its focal length), but astronomical objects are effectively at infinity. When imaging an object at infinity, a spherical mirror causes light from its center to be focused farther away from its surface than that from the edge. (This same principle is why some older macro camera lenses do not work very well for imaging stars.) By using a paraboloidal mirror surface (which in effect has a slight but progressively larger radius of curvature toward its edge), the surfaces of outer zones have a shallow enough angle to bring light to focus at the same point as light from the center.

Astigmatism results when an optical surface has different effective radii across different axes, giving it a shape sort of like the side of an American football.

This distortion is usually so slight that it cannot be detected just by looking at the optics. Moderate astigmatism is visible as an elongation in slightly defocused star images. Orientation of the elongation will shift 90 degrees on either side of focus. When strong astigmatism is present, the elongation will not completely disappear when the image is focused, and a star may look like a plus (+) sign.

Coma is an aberration which occurs in off-axis parts of the image. Coma in a Newtonian telescope produces a teardrop or "v" shaped off-axis star image which flares out in a direction perpendicular to the optical axis. The orientation can be reversed in some Cassegrain telescopes which have coma.

The fact that a telescope has coma does not mean that it is poorly designed or made. The Newtonian is one of the most elegant telescopes ever devised, yet off-axis coma from its comparatively fast paraboloidal mirror is the price one has to pay for getting a sharp axial image and a fast f/ratio at a low cost. Corrector lenses will reduce coma at the expense of introducing a limited amount of new aberrations, but these new aberrations are not a problem for most wide field images.

Other aberrations may be the result of optical compromises which are necessary to provide a good balance between central image quality, off-axis image quality, and cost of manufacture. If such compromises are applied to a conventional telescope, the central star image should at least appear to be symmetrical. If a central star image does not look reasonably symmetrical, it usually indicates that something is amiss in terms of manufacture or optical alignment. The exception is in the case of off-axis telescope designs, since some of these may have noticeable residual aberrations.

Stress on optical components can make a star image appear like anything from a triangle to a completely asymmetrical shape. Most of the time, stress is due to inappropriate mounting of the optics, but poor annealing of optical material can cause a mirror or lens to be internally stressed even if it is just sitting around like a sack of flour. Internally stressed optics tend to be excessively susceptible to disfigurement at different temperatures, making them a real pain in the neck.

Tube currents can have a variety of effects and can cause a star to appear slightly triangular in some cases. You can usually determine if currents are the culprit by looking for what appear to be moving shadowy ripples or funnels in out of focus star images.

Chromatic aberration is common in telescopes having refracting elements. It relates to the refractive index of optical material rather than the accuracy of its surface. Longitudinal chromatic aberration is where a lens brings different colors to focus at different distances from itself, resulting in a blur from out of focus colors when one color is properly focused. It is not uncommon to see a predominantly violet blur around star images produced by doublet refractor telescopes. Lateral chromatic aberration is when the image scale changes according to color, resulting in color fringing perpendicular to the optical axis. Old or simple fisheye camera lenses tend to be more subject to this than telescopes because the outer zones of their front elements refract light at a larger angle.

Field curvature and flare are not always classified as aberrations in the literal sense because they do not influence resolution of the in-focus image; however, they do degrade overall image quality in real world pictures.

Field curvature means that the surface of best focus is curved rather than being flat. Most cameras have flat focal surfaces, so only one radial zone of an image having field curvature will be in perfect focus. Fortunately, field flatteners are available for some telescopes, and many of these can be adapted to telescopes made by other manufacturers.

Flare causes light to be spread over part or all of the image area. The intensity of flare is usually low, but its presence lowers contrast. Dark areas of the image are most affected by flare because the intensity of flare is greater relative to a dark part of the picture. Some flare is caused by reflections from optical surfaces, but good design and antireflection coatings can minimize its effect on refracting optics. Special reflective coatings can also reduce flare and blooming from mirror surfaces.

Stray light or reflections from the inside of an optical housing can also cause flare. All optical housings will cause some flare, but a telescope with poorly designed internal baffles can have an unacceptable degree of it. Good baffle design is particularly important for slow Cassegrain reflector telescopes because stray light getting past its baffle tubes can easily overwhelm off-axis parts of the image. It is amazing how few commercial Cassegrain telescopes are properly baffled. A few that got it about right include the properly baffled Questar 3.5, the Quantum 6, and the Celestron 8.

For photos of subjects like nebulae or open star clusters, correction of field curvature and off-axis aberrations can be even more important than whether or not a telescope resolves the theoretical limit on axis. This is not to say that you should avoid getting a telescope that does resolve to the theoretical limit at the center; most people want to look at planets at some time or another, and a telescope must approach theoretical performance if you are to get the best planetary views possible.

In review, major factors influencing photographic resolution include camera or subject motion, tracking accuracy, film grain, optical quality, diffraction, atmospheric conditions, f/ratio, and image scale versus aperture. Exposure time can affect resolution under some conditions because more effects from undesirable atmospheric conditions can accumulate during a longer exposure.

4.4 Subject Brightness, Aperture, and f/ratio

In astrophotography, it is important to appreciate the difference between recording a dim point source such as a star, or an extended object such as a nebula.

To a large degree, the aperture diameter (not the f/ratio) is what determines how dim a point source object (such as a star) can be imaged in a given time on a given film. This is because most good optical systems will produce a star image (Airy disk) which is smaller than the film grain. The sharpness of most film is limited by the size of the smallest grain unit it can produce, so a dim star will be imaged at least as large as this grain unit regardless of how small the Airy disk is. For a given aperture, roughly the same number of photons will affect the grain unit regardless of the f/ratio, so long as the Airy disk is small enough.

The absolute size of the Airy disk is dependent on f/ratio, with the disk being larger for a slower f/ratio. The f/ratio and corresponding size of the Airy disk do not become relevant in most film photography unless the f/ratio is unusually slow; say about f/22 or slower. (The exact minimum f/ratio is obviously dependent on what film you use.) At this point, the

Airy disk becomes larger than the film grain, making it the limiting factor in terms of resolution. Since a large Airy disk effectively spreads the light from a star out over a larger area, an excessively slow f/ratio can cause point source objects to behave as extended objects in terms of requiring more exposure.

The importance of aperture versus f/ratio can best be appreciated by comparing extremes. An 8 mm f/4 fisheye lens has an aperture of only 2 mm, while a 20 cm f/4 telescope has an aperture of 200 mm. This is 10,000 times more aperture area than the fisheye lens! Even though both systems have an f/ratio of 4, the telescope will capture stars up to 10 magnitudes dimmer than the fisheye lens, given the same exposure time. One way to look at this is to picture how much starlight can fall on the aperture of your optical system. A bigger aperture will obviously have more total starlight falling on it. This is why some people refer to a large aperture telescope as a "light bucket".

For reference, each successive stellar magnitude represents a difference in intensity of about 2.51. This value was selected because five magnitudes equal a one hundred fold difference in intensity. Therefore, a second magnitude star is 2.51 times dimmer than a first magnitude star; a third magnitude star is 6.3 times dimmer than a first magnitude star; a fourth magnitude star is about 16 times dimmer; a fifth magnitude star is about 40 times dimmer; a sixth magnitude star is 100 times dimmer; and so on.

If you are instead photographing extended objects such as nebulae, it is the f/ratio that will determine how bright an object you can image in a given time with a given film. As shown in Chapter 1, an optical system with an f/ratio of 4 will produce an image four times brighter than one with an f/ratio of 8, allowing the exposure time to be shortened by a factor of four. (Possibly even more when the film's reciprocity failure is taken into account.) However, aperture is still important for photographing nebulae because a larger aperture will allow you to get a larger image while retaining a given f/ratio.

Nebulae can be a bit more fussy about exposure than other subjects. Underexposure can keep you from getting any visible images of nebulae because a certain exposure threshold must be reach before a visible image is recorded. Too much exposure can lower contrast by exceeding the sky fog limit. In the case of most dimmer nebulae, the best results are realized when your exposure reaches a point where the sky fog (the background sky brightness or "sky glow") just is barely recorded on the film.

Many deep-sky objects are richly colored. Reflection nebulae are blue or other colors and emission nebulae are typically red or pink. Adequately exposed color photos will reveal these colors even though such color is not observable visually. You can't visually observe the true color of nebulae because of how dim they are. When you look at a dim subject, the rods in your eye (which are sensitive to dim light but not to color) take precedence over the cones (which are sensitive to color) that you utilize in brighter conditions. The concept is easy to comprehend by considering that the color of a flourescent pink jacket is obvious during the day, but if you were to look at the same jacket on a dark night, you would not be able to discern its color very well; it will just appear to be a dusky gray.

Meteors are a hybrid photographic subject because they have a rapid angular motion and a variety of other attributes. The brightest part of a meteor is more or less analogous to a point light source (particularly if a wide angle lens is used to photograph it), while any glow around or behind the brightest part is more like an extended source. Therefore, absolute aperture is important for recording a streaked image from the brightest part of the meteor, but f/ratio is important for recording any associated glow or residual trail. This is why you can often capture dimmer meteors with a normal camera lens than you can with a fisheye lens or all-sky camera having the same f/ratio.

The times when f/ratio can be important in recording the central part of the meteor inlude occasions when the meteor moves at a fast angular rate. Here, a fast f/ratio wide angle lens may capture a dimmer meteor than a longer but slower lens having the same absolute aperture diameter because the wide angle lens compresses the meteor trail into a shorter image on the film.

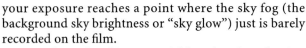

4.5 Tradeoffs

As covered earlier, there is a limit to how much it is practical to increase focal length. Optics aren't perfect, so increasing the focal length too much will just give you a big and fuzzy image. In addition, using longer focal length optics of a given aperture will result in a slower f/ratio, which in turn will dim the image and increase the required exposure time. A longer exposure time also subjects your image to a greater accumulation of poor "seeing" events from turbulence in the atmosphere.

A good medium to large aperture telescope can go a long way toward getting better film images of planets, but atmospheric conditions can be a factor here as well. All of these interrelated matters present somewhat of a conundrum, but the key is to concentrate on gadgets and techniques which are best suited to getting the particular type of picture you want.

4.5.1 Atmospheric Effects versus Aperture and Film Speed

A good large aperture telescope provides better resolution than a small one, but atmospheric conditions seldom allow the large telescope's full resolving power to be realized in photographic images. If the atmosphere did not affect your image, increasing the aperture diameter by a factor of four would also increase the angular resolution of a long focal length photograph by about a factor of four. In the real universe, a planetary photo taken through the four fold larger telescope may only be two or three times sharper than an image taken through the small one.

Atmospheric conditions also work against a small aperture telescope, but not as severely. However, the cumulative effects of seeing can have more effect on photos taken through a small telescope because a longer exposure is required to get a planetary image size comparable to that from the larger telescope. The actual results with each size of telescope will vary according to the type and severity of atmospheric turbulence.

Increasing film speed is a useful way to counteract atmospheric problems to a degree, but there are limits to how fast a film you may want to use unless the seeing is particularly bad. Fast films tend to have a larger, harsher grain structure and less dynamic range and color saturation than slow films. Accordingly, a better image can usually be acquired on appropriate slower film when atmospheric conditions are favorable, provided the telescope's tracking is accurate.

Many planetary features have more obvious changes in color than density, so color saturation can be important in getting as much detail as possible in your image. This was repeatedly demonstrated as I was converting my color images to black and white for this book; some features just disappeared when a raw color image was converted to black and white, so a little image processing was required in order to recover features which were obvious in the color originals.

4.5.2 Wide Field Image Quality versus f/ratio and Film Speed

A fast f/ratio camera lens and fast film will capture a deep sky image in relatively little time, but getting the fastest available optics and film is not always the best policy. A fast camera lens of a given focal length must have a larger diameter, so it will tend to cost more, weigh more, and have more optical aberrations. These aberrations are usually the most obvious at full aperture because a larger part of the optical surface is utilized to image each point in the subject.

Stopping a fast lens down one f/stop or so will most likely improve its performance, permitting it to produce sharper images. If your subject is bright enough, it will still be possible to use a reasonable

Figure 4.2. Image quality in piggyback photos can often be maximized by stopping a camera lens down a modest degree. The left image of Cassiopeia is enlarged from a negative which was exposed for two minutes through a 55 mm f/1.2 lens which was working at its maximum aperture. The right image is a 5 minute exposure taken through the same lens, but at an f/ratio of f/2.

exposure time even when the lens is stopped down a moderate degree. If you are shooting black and white photos, you can try an alternate method. A strong color filter (typically red, orange, or yellow) will usually reduce the visible effects of some lens defects in black and white pictures.

There are circumstances that may require a lens to be stopped down even if it is capable of producing a sharp image wide open. Some lenses may have excessive light fall-off toward the corners of the format when used at full aperture. This will produce a large "hot spot" in the center of your picture. An example of this is the Milky Way photo in Figure 7.3 which was taken with a 105 mm f/2.5 lens.

You can evaluate how much light fall-off a lens has by opening the back and shutter of your camera, then looking at your lens through a corner of the shutter opening. If a large part of the lens aperture appears to be cut off by the edge of an optical element from an off-axis vantage point, stopping down the aperture will probably improve things.

The true working aperture of your optics can usually be determined by looking into the front from a distance. The maximum aperture of a wide angle lens is typically far smaller than its front element. The maximum aperture of a telephoto lens having a focal length at least twice as long as the diagonal of the film format is usually about the same size as the front element.

It is important to point out that while stopping down a camera lens can often improve performance, the same should not be true of a good telescope. It can be practical to stop down a telescope to get better planetary images under bad seeing conditions; but a decent telescope should not have to be stopped down to get a good central image under good seeing conditions. Remember that the theoretical resolution of a large aperture telescope is better than that of a small one. If a telescope has to always be stopped down in order to get a decent central image, it simply is not made or aligned properly!

4.5.3 Film versus Electronic Imaging

Film imaging is the primary subject of this book, but in keeping with the theme of utilizing equipment one may already have, the following section will briefly deal with imaging that does not utilize film.

A few readers may have an electronic imaging device such as an integrating electronic imager, a digital still camera, or a video surveillance camera. These relatively specialized devices are covered in another book in this series.

Electronic imaging through a telescope can also be accomplished with less specialized equipment. You can use a camcorder. Many people already have a camcorder, and those who do can get into some types of astronomical imaging by using it either alone or with a converter lens or telescope.

4.5.3.1 Astro Imaging with a Camcorder

Used alone, most camcorders will allow you to get video of events like planetary conjunctions during morning or evening twilight. A camcorder can also provide a reasonably good image of the moon if it has a long range zoom lens or you use a 2× or stronger tele- photo converter lens. You can even get video of the sun if you use a suitable solar filter in front of your optics. The solar filter must always be attached when the cam- corder is pointed at or near the sun in order to prevent damage to its image sensor.

Camcorders do not offer super long integrations like cooled CCD cameras, but they do offer the advantage of a small sensor size for planetary imaging through a telescope. Deep sky objects are out of reach for virtu- ally all camcorders, but an image intensifier can permit you to image some dim objects on film or video.

Small image sensors can work very well for imaging planets and other sufficiently bright narrow angle objects. This is mostly due to the sensor's small pixels, but the size of the sensor format is also important, given the limited resolution of a video frame. For example, the 4.4 mm width of a popular $\frac{1}{3}$ inch format sensor is about 8 times smaller than the width of the 35 mm film format. The smaller format sensor allows a given field of view to be covered with less focal length. In this case, only $\frac{1}{8}$ as much focal length is required to fill the sensor's format with a subject of interest. A more important feature is that each pixel in a typical small video sensor is smaller than the unit grain size of most photographic film having the same sensitivity. Therefore, a relatively small image on the sensor can provide the same angular resolution as a larger image on film.

A suitable small sensor can make it possible to get a full frame planetary image; something almost unheard of in film imaging with small telescopes. The shorter required focal length and resulting faster f/ratio (given the same aperture) permits the exposure time to be far shorter for the electronic sensor than it would be for a typical film image having the same detail. The shorter exposure in turn lessens the cumulative effects of atmospheric turbulence on the image.

Some camcorders have a digital zoom feature which can enlarge a small image at the center of the sensor enough to fill the entire video screen. Used in moderation, this can provide an even larger image. Too much digital zoom will produce a fuzzy or pixilated image, but even this can be useful for focusing.

4.5.3.2 Video Sensor Formats

It may be worth clearing up a few things about electronic sensor formats. Many video sensors are classified in terms of a fraction of an inch, such as $\frac{1}{2}$ inch, $\frac{1}{3}$ inch, and so on; however, these figures do not actually refer to any dimension on the sensor format. Instead, they are carried over from the days when video cameras used a tube (such as a Vidicon or Newvicon) instead of a solid state sensor such as a CCD. The format size refers to the designation of the video tube. The tube has a larger physical size than the utilized part of its active imaging area, so the tube's specified size and its actual format dimensions are of different values.

In the case of a CCD, the active imaging area is roughly the same size as the utilized part of the imaging area in a video tube sharing the same format designation. Some of today's smaller sensor formats were not around when video tubes were widely used, so smaller format sizes were extrapolated from the existing larger formats. Knowing the actual size of your video sensor can be useful when calculating your focal length or selecting which optics to use. Approximate dimensions for various video sensor formats are shown in Table 4.1.

Table 4.1. Dimensions of video sensor formats

Specified format (inches)	Vertical dimension (mm)	Horizontal dimension (mm)	Diagonal dimension (mm)
1	9.6	12.8	16.0
$\frac{2}{3}$	6.6	8.8	11.0
$\frac{1}{2}$	4.8	6.4	8.0
$\frac{1}{3}$	3.3	4.4	5.5
$\frac{1}{4}$	2.4	3.2	4.0
$\frac{1}{6}$	1.7	2.2	2.8
$\frac{1}{8}$	1.2	1.6	2.0

4.5.3.3 Using a Camcorder with Your Telescope

Most camcorders do not have removable lenses, so to use one with your telescope, you can utilize the afocal method. As covered in an earlier chapter, the afocal method is where you simply point your camera lens (or camcorder lens) into your telescope eyepiece. In some cases, all you need is a second tripod for your camcorder or a bracket to properly position it on your telescope. I usually use a separate tripod because it keeps vibrations from the camcorder from being transferred to the telescope. It also allows me to gain visual access to the telescope eyepiece by simply swiveling the star diagonal or tilting the camcorder tripod back on two of its legs.

By using the afocal method, you can most likely get images of objects such as the moon and brighter planets with your camcorder! The afocal method allows you to adjust the image size with the zoom lens on your camcorder and using different eyepieces on the telescope. Some camcorders only have automatic exposure control. With these, your available exposure control options may be limited to changing the image size (which in turn changes the effective f/ratio), using filters, or covering part of your telescope aperture. The latter is not always desirable because it can reduce resolution when atmospheric seeing conditions are favorable. A video camera with a manual gain control can better allow you to reduce video noise and control the image brightness. You don't want to use a true "manual iris control" for this because stopping down the iris can introduce vignetting in an afocal image.

Afocal video imaging can be very effective. On my first try at afocal planetary imaging with my camcorder, I was able to acquire color images of Saturn which exceeded the quality of any film image I had ever been able to get with the same size of telescope. Obviously, the afocal technique is also applicable to digital still cameras.

The camcorder I use most is a JVC GR-SZ7, a small SVHSC camcorder having a "slow shutter" feature that provides exposures (integrations) up to about one full second. (The feature is entirely electronic so it does not utilize a mechanical shutter.)

A one second exposure provides enough sensitivity to get a relatively large planetary image even with

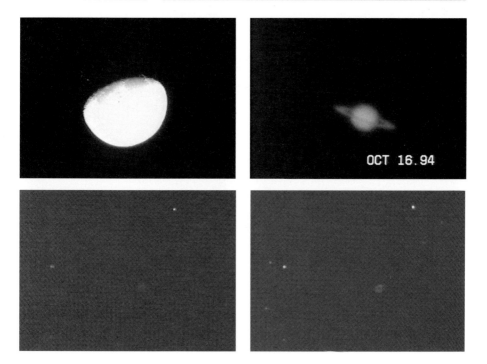

a small telescope. Even better, the camcorder continuously captures one image after another, thereby providing a lot of successive images to select from or combine. The images can be recorded directly to video tape, captured live with a frame grabber, or both. Since the camcorder also records audio, a verbal description of the optical setup and observing conditions can be recorded. It's all quite painless.

If you don't have a frame grabber and want a print of your video, you can use a video printer or just take a picture of your television (TV) screen with your film camera. If you take a picture of your TV, it is best to do so in a darkened room and to use a shutter speed of 1/15 second or longer. Too fast a shutter speed may result in uneven brightness or image only part of the screen area.

There are ups and downs to both film and electronic imaging, but in general, I have found electronic imaging with a camcorder to be more satisfactory for planets than it is for wide field subjects. In spite of the relatively low resolution of a video frame, there are times when there is no substitute for the benefits it can provide. If the subject has a small angular size and the image scale is large enough that the sensor pixels are smaller than about half the spatial resolution of the telescope, the telescope rather than the sensor will be the limiting factor in terms of resolution. In addition, video with sound will capture an astronomical event like the beginning of a total solar eclipse in ways that no still image can.

Chapter 5
Travel and Astrophotography

Using a telescope can be easy if you can just carry or roll it outside or if it is in a permanent observatory; however, few of us have the advantage of living in an area with a dark and unobstructed night sky. Still more of us may not even have a yard of our own in which to set up a telescope. At one time or another, something near our place of residence will probably prevent us from shooting the astrophotos we want from home. Whether the local problem is a tree, a neighbor's light, or a city full of lights, some local factor is bound to make travel necessary.

Getting away from city lights is probably the most common reason astrophotographers travel, but other considerations such as getting an unobstructed horizon below an object of interest, finding an interesting foreground for a twilight photo, or dodging clouds can also bring about the need for travel. At other times one may just want to bring a telescope along on their vacation. In more extreme cases, people travel to other countries in order to observe an eclipse or photograph celestial objects at declinations which are not observable from their home latitude.

Some measures can be taken to minimize observing problems near one's home. These include politely talking to a neighbor about a light they may habitually leave on; requesting that the city put a shield on a nearby street light; making a portable blind to shield you and your telescope from stray light; or installing multiple permanently aligned telescope piers in your yard so you can switch between them to get more sky coverage. Perhaps you could make bird baths that fit the piers so they will look better during the day.

5.1 Packing Equipment for Efficient Transportation and Setup

Travel usually requires that you pack your equipment in the smallest space possible. This can require some disassembly of a telescope or at least removing some of its attachments, so setting up your telescope at a remote site may take longer than it would at home. Part of successful "portable astrophotography" is taking steps to minimize the amount of set up time that is required. After all, you don't want to become so tired from driving and setting up your equipment that you lose interest in taking the astrophotos you traveled so far to shoot!

One way to shorten setup time is to make or acquire gadgets which can be left on the telescope when it is stored in its case. Saving the time it takes to attach a few things to a telescope tube may not seem like much, but it all adds up when you are at a remote site, particularly if it is cold or the wind is blowing. Using bolts with knurled or scalloped knobs instead of conventional heads can improve setup time, but these should only be used in places where mere hand tightening of a bolt will be adequate. Attaching a visual or tactile level indicator to a tripod or using strips of Velcro™ to hold small items can also make it easier to both set up and use your telescope.

A flip mirror attachment is also a time and space saver if its reflector is good enough to allow it to be used instead of a separate star diagonal. This will allow you to leave it on your telescope practically all of the time. A flip mirror can keep you from having to juggle visual and photographic accessories. Various multiple function gadgets can also simplify your system.

Equatorial mounts which have a built-in drive corrector that can run directly from internal or external batteries are becoming more common. These can save a lot of time, space, and weight. Another handy item is a tripod having adjustable length legs. Still more space

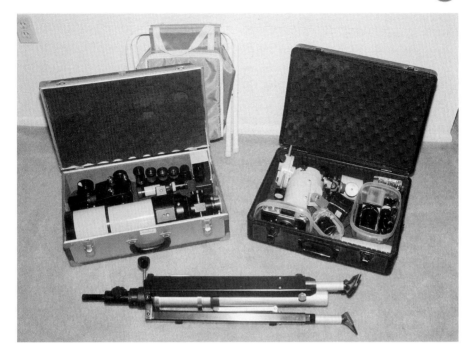

Figure 5.1. My Vernoscope 9.4 cm f/7 refractor and its Aus-Jena equatorial mount have been modified to fit into small cases, yet be set up quickly. Plastic food containers in the cases hold a camera body, guider insert, tools, and a few eyepieces. The plastic bags I normally use to protect other items have been removed in order to show the equipment more clearly. This telescope is shown set up in Chapter 2.

and weight can be saved by using a short tripod which is the right height to allow comfortable observing when you are seated on a light folding chair or camp stool; this provides the added advantage of allowing you to be seated while guiding deep sky photos.

Telescope and accessory cases can be exposed to a lot of dust and dirt when you travel, but keeping your equipment clean is very important. Packing each major piece of your equipment in a plastic bag or putting groups of items in plastic food containers can help keep them clean.

5.2 Light Pollution and Site Selection

"Light pollution" is an increasing problem throughout the world. In some cities, light pollution is so bad that even the full moon will cause little visible difference in the brightness of the night sky. Light pollution from the moon or city lights can affect how dim a subject you can effectively observe or photograph. Fortunately, not all subjects are sensitive to light pollution. For instance,

planets can be observed from either a dark location or from the city. In fact, some of the best planetary views I ever had were from right in Phoenix, Arizona. All I had to do was go out in the back yard and observe!

If you want to observe and photograph nebulae, putting distance between yourself and light pollution sources is the best solution. To get a really dark sky, it is often necessary to put a good 100 kilometers between you and the nearest city of half a million people or more. Obviously, you want to try to have the city behind you when you are facing the objects you want to photograph. The required distance from a city depends in part on the type of light sources the city uses and how well they are shielded.

Some cities use antiquated light fixtures that allow a substantial portion of their light to go directly up into the sky where it will do no good in lighting the street. One would think that cities would wise up because it costs extra tax money to light up the sky. Properly shielded street lights have reflectors to direct the light down where it is needed. Some of these lights are so efficient that they will pay for themselves in a few years through reduced energy costs.

Light pollution rejection (LPR) filters (also called deep sky or nebula filters) are effective for blocking some degree of man made light pollution. They have less effect on moonlight because of its relatively full spectrum nature. Some LPR filters are made specifically for visual observation while others are optimized for photography. Such filters are no substitute for a dark site, but they are certainly better than nothing if you can't get out of the city.

Most LPR filters are designed to work best for emission nebulae, and a good one can even enhance photos of emission nebulae which are taken from dark sites by reducing the size of star images in relation to the nebula.

Figure 5.2. Left: Excessive light pollution can obscure most nebulae. This photo of Orion was taken from a small town, showing that it only takes a few nearby lights to cause a problem. Far fewer features are visible from a large city. (This is a 5-minute exposure at f/2 through a 55 mm lens on Ektachrome 200 film.) Right: A dark site away from artificial lights will reveal many more wonders of the night sky. A darker sky permits more exposure without visibly recording sky glow on your film, resulting in better images of nebulae. (This is a 12-minute exposure at f/2 with the same lens on hypered Ektachrome 400 film. Barnard's loop is the large arc of faint nebulosity left of the belt and sword of Orion.)

LPR filters are a bit less effective for galaxies, since some of the light from a galaxy is of the same wavelength as the light pollution you want to filter out; however, some manufacturers make specific filters which are optimized for galaxies and other selected types of subjects.

If you get a photographic LPR filter, be sure to get a *good* one. I have used some "bargain" LPR filters that did not work so well. They filtered out the light pollution OK, but they also filtered out a lot of the subject I wanted to photograph. In some cases, their transmission at desired wavelengths was so poor that I could not photograph much of anything. Some of my exposures through cheap filters were up to 90 minutes, yet there was no trace of the nebula I was shooting in my pictures! The added expense of wasted film, gasoline, and time easily exceeded the difference in cost between my cheap LPR filter and a good one.

Even if you have an LPR filter, a dark site can make your outing more enjoyable. There is nothing quite like the feeling you get when you can just look up and see the Milky Way (our galaxy, not the candy bar) with the unaided eye.

Going to a dark site is not always as simple as it once was. It used to be that you could just go out alone or in a group and set up a telescope on public land without any hassle. These days, some public land agencies say they want people to pay if they park a car for even a brief time. Others say they want people to pay royalties if they sell pictures they shoot in the area. Still others have very ambiguous and inconsistently enforced policies. Another concern in some areas is safety. With the growing number of whackos out there robbing people and even worse, careful site selection is becoming increasingly important. Such developments can take a lot of the fun out of things.

5.3 Suggested Etiquette at Sites Used by Multiple Observers

It is not unusual for amateur astronomers to gather for a "star party" at a dark site on set nights for purposes of safety, camaraderie, astrophotography, and astronomical observation. Some who live in a large city may go to a lot of trouble to get to a dark site, and most only get the opportunity every month or so. With the opportunity to observe from a dark site being so rare for some people, they understandably do not want others interfering with their astronomical efforts.

Allowing one's eyes to become dark adapted is necessary to get the full benefit of a dark site. The full extent of this dark adaptation can be lost for at least several minutes if someone blasts you with their headlights, lantern, flash unit, or even a flashlight. Astrophotography sessions can also be brought to naught if someone drives too close to your telescope or piggybacked camera with their headlights on or walks up and uses their flashlight to get a better look at your telescope.

If you use a flashlight around other observers, you should use a red one and shine it only on the ground in order to shield observers from its direct light. A conventional flashlight can be made into a red one by simply adding a DEEP red filter. If you find yourself at a star party with only a white light and you can't see without it, you can minimize its effect on others by pointing it straight down and shining it through your hand or closed fingers.

Using care with lights and refraining from taking flash pictures around other observers can make things better for everyone. We would not be happy if someone messed up our night vision or photos with a light during a star party, so it is important to show the same consideration to other observers that we would like them to have toward us; sort of a "golden rule" for astrophotographers. If you want to look at another observer's telescope, don't shine any sort of light on it or stand in front of it unless the observer indicates you can do so. There is nothing quite like the frustration you can feel when some bystander suddenly shines a flashlight into your telescope when you are nearing the end of a one hour or longer guided exposure. My guiding system has a built-in manual shutter, but I once had someone do this to me so suddenly that I had no time to react. Such encounters may cause some serious observers to avoid public events and instead concentrate on observing with their own clique.

Car lights at an observing site are another consideration. Even the light in the cab or trunk of a car is enough to hinder many observers, and simply removing the bulb from the dome light or trunk light is an easy way to keep a car from being a problem.

Headlights are obviously an even worse problem. If you expect to leave a site before most other observers, you should try to park close to the site exit with the front of the car toward the exit. When leaving, one should not just turn on their headlights without giving other observers enough notice that they can at least cover their eyes and any necessary optics. Instead, observers should ask if leaving will affect anyone else, and delay their departure (at least to a reasonable degree) until a better time if it will. Other observers at regularly scheduled public star parties can and should reciprocate by agreeing on several set times throughout the night when people can drive.

When driving through the site, you may want to drive slowly with only your parking lights on if it is safe to do so. It is also a good idea to have your car window open so you can hear people who may be directing you toward the exit. Where dust is a problem, it is particularly important to drive very slowly; sometimes even slower than the speed that a person would walk.

Lights from passing cars do not affect pictures taken through well designed and properly shielded telescopes as much as many people think they do (I have never had a picture spoiled in this way) but not everyone uses a well designed telescope. In addition, piggyback images through fast f/ratio lenses are likely to be affected by car headlights. Therefore, it is best to assume that manmade light sources will adversely affect other observers.

A rarely encountered consideration is the use of infrared flashlights and image intensifiers. The infrared or near infrared light from these flashlights is invisible to the unaided eye, but it can affect some films and image sensors. Therefore, it is best not to shine any sort of light on a telescope (even apparently an unattended one since it could be taking an autoguided photo), because the light could spoil a deep sky photo that is in the process of being taken.

Noise can be another problem for some observers, and recently problematic sources of noise are car alarms and keyless entry systems which make loud chirping noises. I fail to see the advantage of keyless entry chirps which are so loud they can be heard from a block away; all it does is alert the entire neighborhood of when you are coming and going. Car alarms that sound off for no reason are just an annoying modern-day version of the little boy who cried "wolf".

At a star party, some observers may get tired after a long trip to a dark site, and a few take a nap during part of the night. Others may be shooting pictures with autoguiders while they sleep. If a star party lasts all night, some people may try to sleep until at least mid morning before driving back home. Car alarms, high pitch keyless entry chirps, and loud talking can interrupt people's sleep. Car chirps or alarms can do this both because they are loud and because some of them have the same pitch as the electric alarm clocks many people use. People obviously don't like to go to the trouble of getting away from the city only to end up being startled or awakened by some other person's car alarm or loud keyless entry chirp. They can experience that at home!

Noise can also be a problem for observers who are not sleeping. If an observer is really concentrating on setting up their equipment or guiding a photo, being startled by a loud car alarm, horn, or chirp could temporarily turn a normally calm observer into a quivering mass of protoplasm, possibly resulting in a ruined picture or even dropped and broken equipment.

If you have a car which makes loud chirps or other objectionable noise, do whatever you can to keep it quiet. For starters, don't arm its alarm at a star party. A second thing is to use a key to lock and unlock your car if this will keep it from chirping. By being considerate to other observers, you are more likely to be welcome at future organized astronomy outings.

5.4 Astrophotography and Foreign Travel

Astrophotography is an activity which can be enjoyed throughout the world. By traveling to latitudes in the opposite hemisphere, you can gain access to a host of exciting celestial objects that are not visible from home. Travel can also put you under the shadow of the moon for a total solar eclipse or in the right place for other celestial events. Whatever your plans, foreign travel can offer the additional rewards of experiencing exotic sights, interesting people, and other cultures first hand.

Travel can also expose you to various conditions which can be bothersome or harmful, and these conditions can arise from a variety of sources. Some trips can be very pleasant, while others can make you feel as though Murphy's law is the dominant force in nature. The good aspects of a trip need little explanation, but undesirable events can materially affect your situation.

A few potentially undesirable travel issues will be mentioned here. Most of these relate to experiences myself and others I'm acquainted with or heard of have had. By learning from our collective misadventures, you may be able to avoid similar problems. Proper planning is important because it can make a good trip even better than it would be otherwise. Planning can also go a long way toward minimizing the affects of undesirable circumstances, particularly if your plans include contingency options.

There is no way one can anticipate everything which may interfere with an astrophotography expedition. Apart from the all too unpredictable weather, potential interference can arise from microorganisms, insects, and other natural influences. These influences do not have to occur at the time of your planned observations to be a problem. If you contract dysentery or receive a painful sting from an insect prior to your astronomical activities, it can be a major distraction. Therefore, depending on the emphasis of your expedition, it is worthwhile to take precautions like drinking only bottled water, avoiding fresh fruit or salad which may have been washed in untreated local water, and avoiding conditions where you are likely to be exposed to harmful insects, animals, or plants.

When you travel, you can go in a large group, a small group, or even plan your own solo expedition. Large commercial expeditions tailored for astrophotographers or eclipse chasers can offer many advantages, but some can be over-scheduled and relatively expensive. If you make your own travel arrangements or select a tour package that is not specifically associated with a specific theme such as astrophotography, it may be possible to put together a trip for less than half of the cost of going on a targeted commercial expedition.

5.5 Getting Your Equipment Through Airports and Other Terminals

Airports are one of the great bottlenecks of domestic and international travel. You may spend relatively little time in an airport, train station, or other transportation facility, but what happens there can dramatically impact your expedition. It is here that your luggage may be at the most risk or where your film could get X-rayed.

Over a month before you depart, it is a good idea to contact customs in your own country concerning registration of your equipment. Registration is partly for the purpose of proving that you purchased or previously owned your equipment in your own country rather than having purchased it in the country you will be visiting. You may be required to physically bring your equipment and related sales receipts to a port of entry, show them to a customs agent, and fill out some forms. It is also a good idea to check with the proper foreign authorities to see what is required for you to temporarily bring your equipment into the country you will be visiting.

Documentation about your equipment can be useful at security checkpoints in an airport or other port of entry. Many security personnel may be unfamiliar with astronomy related items such as telescopes and equatorial mounts. Here, it may be useful to bring along an astrophoto, an article about astrophotography, or even better, an advertisement or photo which shows what your equipment looks like when it is set up. This will help you demonstrate that your equipment is not a potential security problem.

If the X-ray of your equipment looks unfamiliar to security personnel, you may still have to unpack your carry-on bags for inspection. If you do, you need to be able to repack everything quickly. At such a time, it helps if your carry-on bags are organized and not packed too tightly. I usually pack each major piece of equipment and some padding in a plastic food container before I put it in my baggage. This allows me to remove one piece of equipment for inspection without disturbing too many other items.

It can save time if you wait until you get to your destination before putting film in your camera. This makes it practical to send your empty cameras through an X-ray machine and avoid the inconvenience of having them hand checked. If your film is in a good X-ray bag, you can send it through most X-ray machines without a problem. If the film you have with you is not in an X-ray bag, it can save time if you keep it separate from your baggage as you go through security. Here, it is helpful to keep all of the film together in its own separate container (such as a clear plastic bag), since this usually makes it easier to have it hand checked.

An X-ray machine will not harm your cameras or lenses, and avoiding a hand inspection of these items will save time. What you want to avoid (if possible) is having unprotected film X-rayed, particularly if it has a high ISO number. Your film (particularly after it is exposed) may be the most valuable part of your expedition, and contrary to what some security people may say, X-rays can definitely affect unprotected film!

If you want to be sure you can carry certain baggage with you, it is important to use bags that are small enough. Not all baggage billed as "carry-on" by the manufacturer will be accepted as such by all airlines. Some airlines specify a maximum size as small as 20 cm × 36 cm × 53 cm (8 inches × 14 inches × 21 inches) but many typically allow slightly larger sizes. Bags up to about 24 cm × 36 cm × 57 cm (9.5 inches × 14 inches × 22.5 inches) including handles and other protrusions will easily fit under the seat or in the overhead compartment of most full sized aircraft, but bags only a centimeter larger than this could present a problem.

Most airlines only allow you to carry on two items. A few may even make a fuss even if you have more than one carry-on, so you may want to check with your travel agent and select an airline with reasonable policies. If you take both checked and carry-on bags, it is a good idea to carry on anything that will be essential for at least the first 24 hours of your stay, just in case your checked bags do not arrive on time.

The way you are dressed can materially influence your expedition because it can influence how you are treated at an airport. Sad to say, some airline employees in many parts of the world seem to discriminate against people based on their attire or perceived economic standing. I have noticed that my attire often influences how I am treated at airports both at home and abroad, though it usually does not make as much difference once I get on the airplane. I used to occasionally wear utilitarian clothing like old pants, a white tee shirt, and a photo vest while traveling. At such times, airport personnel have done things like trying to keep me from carrying on as much baggage as other people who were better dressed. On the other hand, when I travel well dressed and clean shaven, I am usually treated well; then some other poor soul who is not as well dressed catches flak from the airline employees. Ironically, one who is relatively well dressed may be able to maintain a lower profile than someone who is not. It is unlikely that this apparent discrimination is purely due to security concerns, since if it were, it would assume that only terrorists and smugglers lack taste when it comes to attire.

Keeping your equipment safe throughout your expedition is another challenge. If your equipment is carried with you, vigilance can help prevent theft. Traveling in a group is useful because group members can look out for each other's luggage. It is also helpful to maintain a low profile by leaving your equipment in your baggage where it is out of sight. If your bags will be just sitting nearby while you wait for your flight, you may want to tie them together with a looped metal cable and a good lock. A cable and lock are also good things to use on a checked bag.

Equipment in checked bags should be well protected. Many people are justifiably apprehensive about how checked baggage is handled. Horror stories about theft and rough handling abound, and no one wants to subject their optical equipment to such risk; therefore, it is usually advisable to pack your optical equipment in your carry-on baggage. This will typically allow you to control how it is handled.

Ground transportation will be required on most expeditions. Whatever country you visit, you will need to get away from city lights in order to take good deep sky astrophotos. Most ports of entry are in a major city, and getting out of the city will most likely involve ground transportation. If you are on a land based eclipse expedition, getting to the path of totality may also require significant ground travel. It is important to consider local infrastructure when selecting your final destination. Some countries have an abundance of four lane highways, while others have only dirt roads in most areas. In many countries, ground transportation to a suitable site can present the greatest logistical challenge of the entire expedition.

5.6 Culture and Your Travel Plans

Some of the greatest and most memorable aspects of your travel experience may be cultural. Most cultural influences tend to be very good, but there are times when the intentional or unintentional actions of fellow travelers or local people may interfere with your plans.

Figure 5.3. Visiting other countries can offer the rewards of meeting remarkable people, seeing interesting sights, and observing astronomical events which are not visible from home.

As with natural influences, some cultural situations can be a problem even if they occur at times other than when you are observing.

Not all influential cultural events are necessarily foreign in origin. For instance, theft of baggage or optical gadgets can occur either in one's own country or abroad. Here, group travel is again advantageous because travel companions can watch each other's baggage.

Cultural factors can sometimes adversely affect travel activities, including ground transportation to an observing site. To a US or west European driver, traffic control conventions in many countries may make driving seem like a fast paced video game. In addition, local traffic may travel on the opposite side of the road that it does in one's home country. Some countries may have few improved roads. Others may require permits to travel from one district to another. Due to these factors, it is often prudent to travel by taxi, public transportation, or chartered vehicles operated by local drivers when possible. If you do choose to drive, you should first be certain that you are familiar with local traffic conventions and that you have the proper license(s) and insurance. In some countries, you can be detained for traffic violations or accidents, even if you think you had "covered all the bases" in your preparations.

5.6.1 Schedule Enough Time for Sleep

Vestiges of one's own culture can be a problem if inappropriately incorporated into group travel plans. One example of this is the tendency of some tour organizers to over-schedule events, resulting in a rushed trip in which there is little time to stop and smell the roses, as it were. Some total solar eclipse expeditions are notorious for this, and in my own opinion, the greatest scheduling mistakes are in allowing too little time for sleep either immediately after arrival in the country, or on the night immediately before the eclipse.

Though most people do not realize that it is happening to them at the time, sleep deprivation can cause a person to become unusually susceptible to making mistakes. It can also keep one from remembering things clearly. I have been amazed at how poorly some people remember a total solar eclipse or other event that they may have spent thousands of dollars to go and see. Almost invariably, most of these same people were members of rushed expeditions that left too little time for sleep. You may spend a lot of money on your expedition to observe and photograph an eclipse, the southern sky, or other subjects. Isn't it worth taking a few steps to get enough sleep and thereby enhance your ability to enjoy and remember the experience?

The local culture at your destination can also influence your efforts, so it is a good idea to determine in advance what your emphasis and objectives are, then plan accordingly. There are potential tradeoffs to consider when deciding how much cultural exposure you want to have. In some areas, increased exposure to a local culture may enhance your efforts and make a trip more enjoyable; in other areas, it may not make any difference; and in still other areas, it could result in interference with your astronomy related efforts.

5.6.2 When Your Emphasis is to Experience a Culture

If you want to really experience a culture beyond what is possible in an organized tour and astrophotography is only an optional activity, then you will obviously want to get out and see the sights and local people as much as possible. This can be particularly rewarding if you have a telescope and you have the opportunity to let local people look through it. In return, you can experience the pleasant nature and good character evident in many of the people you may meet.

Many people have never had the opportunity to look through a decent telescope, so in some cases, looking through a telescope can be a memorable event for them. Saturn has always been a favorite subject among many people who have looked through my telescope, though a telescope is becoming less and less of a novelty in many parts of the world. Even in third world countries, many people already know what Saturn and other planets look like because they have seen planetary images taken by the Hubble space telescope or the Pioneer, Voyager, or Galileo spacecraft.

5.6.3 When Your Emphasis is Astrophotography

If you want to emphasize astrophotography in your travel plans, you may want to consider minimizing your exposure to some local cultures at least during the specific times you want to devote to astrophotography; unless of course you can get together with some local people who are also amateur astronomers!

Most of the time, encounters with local people are pleasant, but there are exceptions in which people may not leave you alone even while you take your photos. It is a good idea to be prepared for such situations. It is even possible that locally influential people will try to exploit you or your telescope against your wishes in order to gain social or political brownie points. In this case, they may even exert pressure on you or local people you have befriended in order to get their way. This type of thing was once tried on a local acquaintance and myself in a South American country, and I later found that the matter may have been orchestrated for the benefit of a local politician.

One is usually less likely to encounter severe problems if traveling as a member of a typical commercial expedition, but such an expedition can also isolate one from truly experiencing the local culture. Even on my least favorable trips, I had at least some opportunity to experience the good and noble character evident in some remarkable local people. In a sense, I would not trade such experiences for anything.

5.6.4 Human Rights Practices and Your Safety

The extremes of what can happen abroad can be quite severe, and the way local people are treated can sometimes be an indication of how you as a traveler could be treated in a worst case scenario. Some potential problems may be the result of crime, while others may be the result of local laws or other government sanctioned activity.

Relatively extreme (but typically rare) threats can include armed robbery, abductions, or worse. Such problems can occasionally arise out of the blue, but more often they may be associated with terrorism, drug trafficking, corrupt officials, intense socioeconomic conflict, ethnic groups which exhibit warlike tendencies, or even outright armed conflict. Still other problems (particularly those associated with terrorist bombs) don't even have to involve an encounter with a person.

Other problems can arise from misunderstandings or local laws a traveler may be unfamiliar with. In some countries, one can get in trouble for using a foreign currency. In others, photographing or showing disrespect to objects locally held in reverence can cause problems. In still others, saying disrespectful things about government or religious leaders can get you thrown in jail or even worse.

Some countries have a policy of only arresting or imprisoning people who violate written local laws. Others may be more arbitrary. The "rights" we in most western countries take for granted may not exist in certain other countries. Even basic rights like the right to a trial (or the presumption of innocence prior to a trial) may not exist. Therefore, it is important to be informed of local circumstances in advance, particularly if you will not be traveling in a large organized group.

It is a particularly good idea to avoid volatile areas, particularly if you travel alone. If you are confronted by the wrong people, possible theft of your equipment may be the least of your worries. Self-defense at such a time is not particularly effective unless you are as slick as Chuck Norris at getting yourself out of a pickle with your bare hands. Many countries have strict gun control laws which have the effect of allowing only criminals to carry weapons, and you don't want to risk carrying a locally banned weapon for obvious reasons.

Before you travel, check with reliable sources about current safety conditions at your intended destination. Circumstances can change rapidly in some parts of the world and your travel agent is not likely to have all the information you may need. Some good sources for information about various countries are on the Internet. Country reports by the US Department of State (published on the Internet) are a reasonably good source of preliminary information, though they may be a bit soft on some countries. Human rights organizations are another source to consider.

There is no way to guarantee absolute safety, but having some reliable information is better than not having any at all. Avoiding a problem is always easier than getting out of one, and you can't necessarily count on your own country's embassy to bail you out if there are problems.

The good news is that bad cultural experiences are typically in the minority for most travelers. Whatever your experience may be on a foreign expedition, there is much to be gained by careful planning which includes preparing a checklist for the items you bring along and for your astrophotography procedures. In preparing these checklists, it is a good idea to practice setting up your equipment, using it, and packing it; and to do this well in advance of your trip. You will be amazed at how valuable such preparation will be, and you will appreciate it a lot when you first try to set up and use your equipment in an unfamiliar area.

Having your procedures down will also help compensate for conditions in which you may have reduced performance; whether the cause is jet lag, sleep deprivation, dysentery, biting insects, or a higher altitude or different temperature range than you are used to. By thinking ahead, you can overcome a lot of potential problems and maximize enjoyment of your travel experience.

Whatever country you are in, the night sky is there for you; and it's tax free!

Chapter 6
Astrophotography Without a Sidereal Drive

Previous chapters have covered basic astrophotography, how to calculate exposure values, what you can realistically expect from various optics, and suggestions for travel and astrophotography. This chapter covers a few techniques for shooting astrophotos with modest equipment. Later chapters will cover more advanced topics.

In addition to providing information about various techniques, this and each of the following chapters include a modest gallery of astrophotos and related exposure data. These photos show the level of results you can typically expect from a given amount of effort when using various camera lenses and telescopes. You can use this exposure data to glean information about exposure values or the desired focal length for a given subject. All (or nearly all in the case of illustrations having an aspect ratio different from the film format) of the frame is reproduced in each illustration except in cases where a different enlargement factor is mentioned in the caption.

You can probably get good astrophotos by literally following procedures suggested in this book, but you may get even better results if you improvise. By tailoring your procedures to match the specific equipment you have, you can maximize its capability.

6.1 Astrophotography Without Tracking

As covered in a previous chapter, celestial objects appear to move across the sky as the earth rotates. This

Figure 6.1. Upper left: A fixed camera with a 35 mm lens captured these star trails behind Colorado's Longs Peak. The exposure was about 15 minutes at f/2 on Ektachrome 200 slide film. The mountains are illuminated by a combination of the rising last quarter moon and light pollution from "front range" cities toward the east. Lights from an airplane caused the dim horizontal streak near the bottom of the picture. Upper right: The moon is bright enough to photograph without tracking. This $\frac{1}{8}$ second exposure on ISO 100 film was taken with an inexpensive 400 mm f/6.3 lens and two stacked 2× tableconverters. Lower left: Events like this 1984 conjunction of the moon and Venus can be photographed with very modest equipment. This photo was taken from a fixed tripod with a 50-250 mm f/4-5.6 Tokina Zoom lens working at a focal length of 120 mm. The exposure was one f/stop less than what the camera's light meter indicated for the sky. Lower right: Even a recognizable image of the dim Milky Way can be photographed from a regular tripod if a fast lens and fast film are used. The tripod was flexed to "track" this 45 second exposure through a 55 mm f/1.2 lens on ISO 400 black and white film. This was one of my very first "deep sky" photos. It was taken in 1979, before I owned an equatorial mount. Today's fast films will provide better results with the same equipment.

motion will result in a streaked image if you take a long exposure from a fixed tripod. To prevent this streaking, most astrophotographers use an equatorial mount and sidereal drive to track celestial objects. Such a mount can easily cost more than a good camera. Fortunately, you can still photograph some celestial objects even if you don't have an equatorial mount and sidereal drive.

Some streaking can be tolerated in an exposure so long as the streaks are short enough that they are not obvious in the image. This requirement for minimal streaking will limit the length of your exposure, which will in turn determine which objects you can adequately photograph without any tracking.

Most constellations, conjunctions, occultations, and eclipses can be photographed with a fixed camera, a normal or telephoto lens, and fast film. In fact, some really good conjunction shots can be taken with a "normal" camera lens. For added interest, the foreground of a conjunction photo taken during twilight can include at least a silhouette of some sort of familiar terrestrial subject.

Even the most basic astronomical telescope can be used with a camera adapter or the afocal method to photograph phases of the moon or partial phases of a lunar eclipse. With a proper front solar filter or projection screen arrangement, you can also photograph the sun. This can include full solar disk shots, close ups of sunspots, or partial phases of a solar eclipse. Rare total solar eclipses can be photographed with almost any telephoto camera lens or telescope, and some aspects of these events (such as the effects of the lunar umbra on the sky) can even be photographed with a wide angle lens.

6.1.1 Untracked Astrophotography with a Camera Lens

The most basic of astrophotography equipment is a simple camera with a normal lens. Such basic equipment can allow you to photograph a surprising array of objects, particularly if the camera has manual exposure settings. If you also use a tripod or other rigid camera support, you will usually get more consistent results.

Constellations and some other wide field subjects can be photographed without any tracking because their brighter stars require only a few seconds of exposure on fast film if a fast camera lens is used. Wide field pictures of other subjects including the moon, a solar eclipse, a planetary conjunction, a bright comet, or an aurora display can also be photographed without tracking.

Figure 6.2. A bright comet can be photographed even from a major city. German Morales, director of Astronomia Sigma Octante, used a tripod mounted camera and 55 mm lens to capture comet Hale-Bopp over the lights of his home city of Cochabamba, Bolivia. The twilight exposure is 6 seconds at f/1.2 on ISO 400 film. Photo Courtesy of Germán Morales Chávez.

Some objects are bright enough to photograph even through twilight or moonlight. The subdued brightness of the twilight sky makes it possible for you to get an image which includes silhouettes of familiar terrestrial objects as well as some celestial objects. Shooting this type of picture can be as easy as taking a light meter reading of the sky, setting the camera to provide one or two f/stops less exposure than the meter says, and taking a picture.

For a given subject, film and lens, image brightness will depend on your exposure time, with a longer exposure providing a brighter image. If the exposure with a fixed camera is too long, visible streaking will result. The minimum resolution you need typically depends on the type of subject and the size of print or display you want to have. Chapter 2 covered how much streaking occurs in a given exposure time and suggested some acceptable tolerances.

Some automatic "point and shoot" cameras can be used to take photographs which include astronomical objects, but if you use one, it is best to try and keep its flash from going off. A flash will not make any difference in the exposure of a celestial object due to the distance involved, and a flash may even overexpose nearby terrestrial objects or bother other observers who may be in the immediate area. In addition, the flash in a point and shoot camera may be coupled to the camera's automatic exposure circuitry, and this in turn may keep the shutter from opening long enough to take a good twilight shot when the flash is used.

Many automatic cameras have a switch to disable the flash, and some of them are capable of exposures of up to $\frac{1}{4}$ second or more when the flash is off. If your camera will shoot a relatively long exposure, you should stabilize it with something like a tripod or sand bag. Other cameras may not have any provision for shooting in dim light without flash. Cameras that don't allow you to disable the flash usually won't work very well for most astrophotography.

If you want to illuminate foreground objects with flash, it is best to do so manually with a separate and relatively weak flash unit. A manual flash unit (particularly one with adjustable brightness or provision for attaching neutral density filters) can provide a lot of additional flexibility. You can get even more control over foreground lighting if you fire your flash at each of several selected objects during a long exposure. Another option is "painting" foreground objects with a flashlight. You can add still more effects by using color filters with your light source. If you use a flash or other light source to illuminate foreground objects, try to do it well away from other observers so the light will not bother them.

Some types of astrophotos can actually benefit from the streaking caused by motion of celestial objects. One example of this is a star trail photo. You can get an interesting picture of concentric star trails if you point your camera and normal or wide angle lens at the celestial pole and take an exposure of half an hour or more. Star trails of other parts of the sky can be equally interesting. Meteors can be captured in a star trail picture just as well as they can in a tracked exposure.

Motion of celestial objects can also be used to advantage in taking a sequence of eclipse or conjunction images. You can obtain an interesting sequence of eclipse images on a single frame of film by simply taking multiple exposures. The moon and sun move their own diameter across the sky about every two minutes, so taking one exposure every four minutes will provide a full lunar or solar diameter of sky between each image of a lunar or solar eclipse.

Sequence pictures are easiest to get if your camera will shoot multiple exposures. If it will not, you can still get a sequence picture with a manual camera during the night. Just put a lens cap on your lens, set the camera shutter on "B", then use a locking cable release to lock open the shutter. If the subject is relatively dim, all you have to do is temporarily remove your lens cap for each exposure. If the subject is so bright that the exposure has to be shorter than a second or so, you can

Figure 6.3. Top: Almost any lens can be used to photograph star trails. This 30 minute exposure was taken toward the north with a 16 mm f/2.8 fisheye lens. A substantially longer exposure will produce longer concentric streaks for a more pronounced effect, but subtle details such as the glow of the Milky Way will become much more blurred. Lower left: A lunar eclipse sequence can be photographed with relatively modest optics. This multiple exposure of the moon entering the earth's umbra during the 6 September, 1979 total lunar eclipse was shot with a 203 mm f/7.7 Ektar lens on Polaroid PN (positive-negative) film. The camera was on a fixed tripod and one exposure was taken every 5 minutes. This 1.5x enlargement encompasses a relatively small section of the 4 × 5 inch film format. Lower right: If local weather won't allow you to see celestial objects, make the weather your subject! This 3 minute exposure of lightning was captured on ISO 64 color slide film with the same 35 mm f/2 lens I use for some of my astrophotography.

use a black card in front of the camera as a "shutter". Here, you put a sufficiently large black card just in front of your camera, remove the lens cap, quickly move the card to one side and then back in front of the lens to make your exposure, then replace the lens cap.

If the exposure needs to be shorter than what you can manage with the card, you can instead make a large black card having a one centimeter or wider slit at or near the middle. The procedure with a slit card is the same as for the black card, except that you always keep the card in front of the camera when the lens cap is off. Instead of moving the card far enough to uncover your lens, you move it so the slit passes in front of the lens. You can calculate your equivalent "shutter speed" provided by the slit by estimating how many slit widths the card moves per second.

The slit shutter method is recommended mostly for wide field images due to diffraction effects from the slit. If you use a slit shutter in front of a telescope and its slit width is much smaller than the telescope aperture, the resolution (at least in the direction perpendicular to the slit) will be reduced to the theoretical limit of an aperture about the same size as your slit. This is not the result you would want if you are shooting close ups of planets, for example.

If you want to get a sequence of the moon moving in relation to other celestial objects, you can use its relative motion as a reference in determining how much time to allow between exposures. The rate of relative lunar motion is dependent on the moon's position in the sky. In general, when the moon is rising or setting, it appears to move eastward (in relation to the stars) at a rate about its own diameter in just over an hour, while it can take close to two hours to move its own diameter when it is near the meridian. The exact rate of relative lunar motion is also dependent on other conditions, including the distance between the earth and moon. The moon's orbit is not perfectly circular, and it moves slower when it is at a greater distance from earth.

The exact lunar rate can be influenced significantly by the latitude of the observer, with an observer near the equator experiencing the slowest apparent motion of the moon. The reduction in apparent lunar speed is due to the earth's rotation carrying the observer in the same general direction as the moon's orbital motion. This effect is also partly responsible for the increased duration of the total phase of a solar eclipse at sites where the sun is near the meridian during totality.

When shooting a sequence of photos, it is important to bear in mind that the moon, planets, and stars will move across the sky at a rate many times faster than their relative rate of motion between each other. Therefore, you can still get a good multiple image sequence shot of a subject like a conjunction of the moon and Venus by taking a picture from a fixed tripod every few minutes. The position of the moon relative to Venus will not change all that much between each shot, but each image in the series will show the gradual change. More interesting close up sequence shots of the moon or planets can be taken by tracking the stars and taking a shot every couple of hours or so.

6.1.2 Untracked Photography Through a Telescope

Your camera and telescope can be used to photograph a wide range of subjects. Some telescopes are marketed for terrestrial use and others are marketed for astronomy, but there is nothing keeping you from using a terrestrial telescope for astronomy or an astronomical telescope for terrestrial observation and photography.

Small but high quality refracting and Cassegrain telescopes can be used to photograph many different terrestrial and celestial subjects. You can use them to photograph the moon, sporting events, car races, air shows, wildlife, and even insects that look like something tabloids may try to pass off as aliens.

Figure 6.4. A telescope can be used to photograph more than just celestial objects. Upper left: A 10 cm f/8 refractor captures the "Blue Angels' at an air show. Upper right: The same optics were used to photograph this elk in velvet from about 50 meters away. Lower left: An 8.9 cm f/16 Maksutov–Cassegrain telescope catches an airplane as it nearly flies in front of the moon. Lower right: The close focusing capability of the same 8.9 cm telescope permits it to spy on a praying mantis from a distance of 3 metres.

Bright celestial objects like the moon and inner planets can be photographed through a telescope even if they are not tracked. Even a modest telescope will usually permit you to get a good picture of the moon, and you can get a larger image by using a teleconverter or Barlow lens. You can add interest to your picture by photographing the full moon as it rises above an interesting landscape in the evening, or as it sets in the morning. You can also get interesting shots of the crescent moon as it rises or sets.

You can get relatively large lunar and planetary images by using the afocal method. This will even allow you to use a fixed lens camera to take pictures through a telescope. For review, the afocal method is where you point your camera and lens (with the lens aperture wide open) into your telescope's eyepiece. You can even use an automatic "point and shoot" film camera for some types of afocal photography, but a camera having manual focus and exposure control will give you a lot more flexibility. If you have an interchangeable lens camera which can be attached directly to your telescope, you can effectively photograph even more objects.

If you lack any telescope at all, you can try using the afocal method with a pair of binoculars. This may cause vignetting (darkened corners) on the 35 mm format, but it will at least provide a larger image than your normal lens. Afocal imaging through binoculars and other optics can also be accomplished with most digital cameras and video camcorders.

Figure 6.5. Left: A 1/60 second exposure on ISO 100 film through a 10.2 cm f/15 refractor provides a stunning image of distant pine trees silhouetted by the rising moon. This image is enlarged about 2.2× and includes about 80 percent of the format. Right: The young crescent moon occults a few stars near the Pleiades during evening twilight. This untracked 5 second exposure on ISO 1600 color negative film was taken through a 20 cm f/3.7 Newtonian telescope on a "Dobsonian" alt-azimuth mount. (This telescope is shown in Chapter 2.)

6.2 Simple and Effective Short Duration Tracking Methods

You can get better images of most celestial objects if you track them during your exposure. Many astrophotographers shoot tracked photos by using an equatorial mount such as that supplied with some moderately priced to expensive telescopes. If you already have an equatorial mount and sidereal drive, all you have to do is polar align it, start the drive motor, and attach your camera and lens to it in a way that still allows it to be balanced well enough to track accurately. Such a mount will usually allow you to take piggyback photos with your camera and just about any lens of small to moderate size and focal length. You can use the telescope and a cross hair eyepiece to verify proper tracking. This type of "guiding" is particularly important when you use a relatively long focal length camera lens.

If you do not have an equatorial mount, you can use many other gadgets and methods to take tracked exposures of modest duration. Your camera does not really have to be on an equatorial mount for exposures of a few seconds; all you have to do is make it "think" it is equatorially mounted for the duration of your exposure. You only have to track an object during your actual exposure; it does not matter what your optics and mount are doing just before or after the time you shoot your picture.

6.2.1 Tracking Without an Equatorial Mount

A tripod will hold your camera still in relation to the earth, but it won't allow you to track celestial objects unless you use it in unusual ways. One way to track celestial objects for up to several seconds is to flex the tripod legs a little by applying gentle but gradually increasing pressure in one direction. Here, a cheap tripod can actually be used to some advantage!

Adequate pressure can typically be applied by hand, and you can verify proper tracking if you have a small finder scope or guide scope and a reticle eyepiece. This type of tracking method is best for relatively short exposures with a normal or wide angle lens.

Longer tracking times can be provided if you orient your tripod head so one axis of its motion points toward the celestial pole. Such an orientation can usually be obtained with a tripod having either a side arm or legs which can independently swing out to a completely sideways position. A tripod having an appropriately tilted axis can be tracked by hand if the head has a sufficiently long handle and you use a guide scope having a reticle. Any sufficiently rigid item having an axis which points toward the celestial pole can essentially be used as an equatorial mount and potentially provide tracking for several minutes if it can be moved smoothly enough.

Easier tracking can be provided by using a slow motion tripod head or alt-azimuth mount which is tilted to the appropriate angle. To track, just turn the slow motion control knob which rotates the tilted head in the equivalent of right ascension. If you know the correct rate to turn the slow motion control, you may be able to adequately track a wide angle exposure for up to several minutes. Still better tracking can be provided by a "barn door" mount such as the one briefly covered later in this chapter.

Whatever tracking method you use, it is helpful to utilize a guide scope and something like a black card or remotely operated shutter plate which can be used in front of your camera lens. If the shutter plate covers only your camera lens, you will still be able to see a guide star in your guide scope when you interrupt your exposure. This in turn will allow you to take breaks when you guide during a long exposure without losing track of your guide star.

Another way to interrupt the exposure is to use a loose fitting lens cap. A tightly fitting clip on cap is not recommended for interrupting exposures because it can be difficult to remove without moving the camera or introducing excessive vibration. If you do not have a loose fitting lens cap, you can make one from cardboard or something like the end of a salt can or mailing tube. Another useful astrophotography accessory is a lens hood, since this can help prevent flare from passing cars or other local light sources.

6.2.2 Tracking With a Nonmotorized Equatorial Mount

Some inexpensive to medium priced telescopes are sold with equatorial mounts that have no drive motor, but most of these have a manual right ascension slow motion control. The add-on motor for some of these mounts can cost up to half as much as the purchase price of the entire telescope and mount, so many people don't get the motor. It is usually best to forgo buying an expensive motor for a wimpy mount and instead save up for when you can replace the telescope and mount with a more robust unit which has a built-in motor drive.

In the meantime, you can manually track your telescope by using the manual right ascension slow motion control your mount may already have. Some mounts have smooth enough manual right ascension slow motion controls to make it possible to manually track a photo at the sidereal rate for a few seconds or maybe even minutes. Mounts that lack precise slow motion controls can still be tracked manually for at least a short period of time, but doing so can be more tedious. Another possibility is to find a suitable and relatively inexpensive clock motor or surplus drive motor having the correct rate and adapt it to your telescope mount.

If your exposure is only a second or two (as it may be for a lunar close up or planetary image), you may be able to track adequately by positioning a tripod immediately under or next to your camera. If the tripod head is tilted in a direction that will allow the telescope to track only in right ascension when you slide the camera along its top, you should be able to get a well tracked exposure.

This type of tracking may be easiest to do when an object is on the meridian, since simply moving the camera sideways will allow you to track objects on the meridian which are located well below the zenith. When a subject is on the meridian, you may even be able to do some tracking with a telescope that is attached to an alt-azimuth mount or tripod head, since you need only move the mount in azimuth to track an object that is on the meridian and a sufficiently large angle below the zenith.

You can also track a picture by referencing your eyepiece or star diagonal to a separate tripod head if you are using the afocal method. If you have a steady hand,

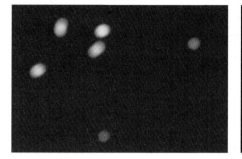

Figure 6.6. Left: These images of Jupiter were "tracked" by moving the eyepiece end of a 10.2 cm f/15 refractor by hand. I just rolled my finger on a nearby tripod so that it slowly pushed the eyepiece to one side. The afocal method was used with a 16 mm eyepiece and 55 mm camera lens to provide a focal length of about 5000 mm. These Kodachrome 64 exposures range from 1 to 6 seconds at about f/50. A black card between the eyepiece and camera lens was used as the shutter, facilitating multiple exposures on a single frame on film. The image is enlarged about 3.5x and includes less than half the film format. Right: This 15x enlargement provides a close up view of a two second exposure from the left photo.

a useful option for tracking with the afocal method is to just grab the eyepiece with a couple of fingers, reference another finger to your tripod or camera lens, and manually move the eyepiece. This is most practical if your telescope has a long tube. You can get a feel for the correct tracking rate by simply looking in your camera viewfinder as you move the eyepiece. In doing this, it is helpful to reference the image to the edge of some feature near the center of your camera's focusing screen. This will allow you to verify that the subject is remaining at a fixed location on the focal surface.

Another way to track when using the afocal method is to make a simple tube from plastic or other pipe, then drill and tap a hole which will accept a machine screw in the side and near one end of the tube. The other end of the tube should fit either in or around the front of your camera lens. The tube can be made to fit if you use a pipe of large enough diameter that one end will fit around your camera lens. This will allow it to be held in place with Nylon set screws. Another attachment method is to use pipe of a size that its end will fit (or can be made to fit) into the front of a filter step up ring. This will allow you to simply screw the tube unit into your lens filter threads.

Tracking is accomplished by orienting the tube so the screw in its side hole will bear against the side of the eyepiece in a direction that will move the entire telescope west in right ascension. If you care about the finish of your eyepiece barrel, you should use a soft point screw and maybe even wrap some tape, plastic or thin cardboard around the part of your eyepiece the screw will bear against. Tracking accuracy can be improved if you add two additional screws to the side of the tube, 90 degrees to either side of the tracking screw. These can be adjusted so their ends just barely clear the sides of the eyepiece barrel and thereby better constrain it to move in the desired direction.

Still another way of briefly tracking with the afocal method involves pivoting the camera and allowing the telescope to remain stationary. This can be a bit more tricky than the other methods because it is likely to introduce chromatic aberration, but it at least reduces the risk of vibrating the telescope. Many other techniques can be employed with simple and inexpensive equipment to track objects for short periods of time, but you may want to purchase or make an equatorial mount of some kind if you plan to take a lot of long exposures which require accurate tracking.

6.3 The "Barn Door" Mount

There are many things to consider when selecting or building a tracking mount for astrophotography. These include stability, tracking accuracy, tracking duration, portability, and ease of setup, alignment, and use. Fortunately, even simple home made equipment can be capable of tracking celestial objects for various periods of time. For most practical purposes, a good home made tracker is an equatorial mount because it can facilitate many of the same results you would get by using a more conventional mount.

Some of the most useful home made tracking systems include those based on the "barn door" mount principle. Simple versions of these can be made from two boards, a hinge, and other inexpensive parts. The hinge makes up the polar axis and a long machine screw or threaded rod is used as a drive screw to control the tracking rate. The drive screw is typically oriented perpendicular to the bottom board and screwed into a fixed nut in the same board. The top board usually has a hard bearing surface which the end of the drive screw pushes against. Turning the drive screw provides tracking by gradually increasing the angle between the mount's two boards. Other variations include a curved bolt which is driven with a rotating nut.

Where a drive screw is used to effect tracking, the drive rate is dependent on: (1) the distance from the hinge to the drive screw; (2) the thread pitch of the drive screw; and (3) the rotation rate of the drive screw or drive nut. Many who turn the screw by hand design their mount in such a way that turning the drive screw at one revolution per minute will provide the correct drive rate. If the mount is motorized, there is no need for the rotation rate to be an exact number of minutes or seconds, so long as it is optimized for the motor's characteristics.

To get the correct tracking rate, you first need to know how far your mount will move with each rotation of the drive screw. If you want to design your mount to have the right tracking rate when the drive screw is spun at a predetermined rate, you first multiply the rotation rate you want in revolutions per minute by the 1440 minute solar day (if your primary goal is to track the sun); or the 1436 minute sidereal day (if you want to track other stars, nebulae, and the like). Once you know this ratio, you can use it to determine the radius a given drive screw must be from the hinge axis of your mount.

For example, say you want to turn a 20 thread per inch drive screw at 1 rpm to track at sidereal rate. (This example is in imperial units because threaded rod of 20 thread per inch pitch is relatively common.)

1 rpm × 1436 = 1436 revolutions of drive screw per day

The correct drive radius will be that in which the circumference has the same ratio to the thread pitch:

1436 × 0.05 inch thread pitch = 71.8 inches circumference.

To get the radius, divide the answer by 2 × pi:

$$\frac{71.8}{6.283} = 11.428 \text{ inches radius.}$$

Therefore, the 20 tpi drive screw should be 11.428 inches (290.3 mm) from the hinge axis.

A barn door mount can "open the door" to a wide range of photographic subjects that are not otherwise accessible. Even better, you can enhance the basic design, making it useful for more than just short exposures with normal and wide angle lenses. In fact, an appropriately enhanced barn door mount can actually provide a more precise sidereal drive rate than many commercial telescope mounts! This accuracy results in part from the relatively large radius between the hinge (or right ascension bearing) and the drive screw. Using a guide scope can improve things even more, since this will facilitate correction of the tracking rate during an exposure.

A very effective enhanced barn door mount was built some time ago by David L. Charles (my brother) of Loveland, Colorado. His mount uses a cam as the bearing surface for the end of the drive screw. This cam corrects tangent error in the drive and facilitates accurate tracking for up to about an hour. Other enhancements include a built-in illuminated clock, a polar alignment scope, quartz controlled stepping motor for the right ascension drive, motorized cross-axis drive (for pseudo-declination correction), provision for mounting a guide scope, and a dual axis remote hand control. All of these features have allowed David to take guided deep sky exposures of up to an hour with lenses of up to 600 mm focal length! What is even more amazing is that he built his entire enhanced barn door tracking system for less than US $75.00. Details about his mount were published in the *Proceedings of the 1988 Riverside Telescope Makers Conference*, where he previously won an award for it. In addition to his own material on the subject, he recommends the April 1975 *Sky and Telescope* article by G. Y. Haig.

A good equatorial mount and sidereal drive will allow you to photograph a wider variety of objects through your telescope and camera lenses. Even a basic equatorial mount and sidereal drive will, if well made and properly polar aligned, provide tracking that is

Figure 6.7. Left: David L. Charles's barn door mount is shown set up for guided astrophotography with a 600 mm lens! A 6cm f/11.7 "department store refractor" is used for a guide scope. The hand paddle on the ground controls tracking in two axes. This view is from the northwest. Right: David Charles used his enhanced barn door mount to take this 30 minute exposure of central Orion with a 350 mm f /5.5 lens on ISO 1600 color negative film. The nebula M42 is to the lower left. Near the top of the picture, the dark "horsehead nebula" appears as a small bump in the nebulosity just left of the brightest star. Both photos courtesy of David L. Charles. All Rights Reserved.

Figure 6.8. A tripod mounted camera with a lens of sufficiently wide angle can capture a total solar eclipse and the horizon in a single picture! This photo of the 3 November, 1994 total solar eclipse is a 1 second exposure through a Vivitar 20 mm f/3.8 lens on ISO 100 color negative film.

accurate enough to shoot unguided deep-sky pictures with normal and wide angle lenses. A good mount and drive will also track well enough to allow you to get better close up images of the moon and planets when you use eyepiece projection or the afocal method. A sidereal drive used in conjunction with a drive rate corrector and guiding system will put even more subjects within your reach.

Chapter 7
Astrophotography With a Sidereal Drive

Previous sections covered what you can photograph with simple equipment and little or no tracking capability. This section covers what you can photograph if you have a motorized equatorial mount or another good way to track celestial objects for an extended period of time. A good tracking mount will permit you to photograph all of the same objects you could without one, plus many more. It will also make it easier to compose your picture when you are using a telescope or telephoto lens because the subject won't keep drifting out of the picture.

Piggyback photography with your camera lens is a relatively easy way to get really good photos of the night sky, including constellations, the Milky Way, and even some larger comets and nebulae. Even a photo taken through a wide angle lens can adequately show some larger objects. For example, the top photo in Fig. 7.1 covers more than half of the sky, yet has enough resolution and image scale to reveal a few individual deep sky objects.

Tracked photography through a telescope opens the door to close up images of the moon and planets, and appropriate guiding accessories and techniques make it possible get reasonably detailed photos of interesting subjects, including star clusters, galaxies, and emission nebulae.

Many astrophotographers who take long deep sky exposures through their camera lenses or telescope tend to use a relatively conventional equatorial mount having a sidereal drive. An equatorial mount and sidereal drive will provide a way to photograph many different objects, but a stable one with an accurate drive can cost more

Figure 7.1. Top: A fisheye lens can capture a vast part of the sky. This 10 minute exposure on Ektachrome 200 slide film was taken with a late 1970s era 16 mm f/2.8 Minolta Rokkor-X fisheye lens. It covers all the way from Orion (left) to Cassiopeia (upper right), yet has enough resolution and image scale to reveal the Pleiades (center), the California nebula (just above center), and M31, the great galaxy in Andromeda (toward right). The streaks at lower right are from residential lights near the horizon. Lower left: Accurate tracking is essential for deep sky astrophotography. This sharp enlargement of the Whirlpool galaxy (M51) would not have been possible without accurate tracking. The image is enlarged 12×, which is about the same enlargement required for a 28 × 36 cm print. The exposure is 6 minutes through a 200 mm f/2.2 lens on ISO 1600 color negative film. Lower right: Eyepiece projection was utilized with a 20 cm f/10 Schmidt-Cassegrain telescope to provide an effective focal length of about 24,000 mm for this image of the lunar crater Copernicus. The 6 mm eyepiece used for projection had to be only a few centimeters from the focal plane to provide a 12 × increase in effective focal length. The exposure was 2 seconds at f/120 on ISO 1000 slide film.

than some want to invest. Alternatives to commercial equatorial mounts include something like the barn door mount mentioned in the previous chapter.

Factors such as tracking accuracy, the weight capacity of your tracking mount, the brightness or angular size of your subject, and the quality of image you need will all determine which objects you can effectively photograph. Tracked wide angle images are usually the easiest to obtain because the tracking does not have to be as accurate as it would for a longer focal length optical system.

The effectiveness of your mount's sidereal drive can be improved if you use a guide scope with a cross hair reticle eyepiece (preferably one with an illuminated reticle) and some method of making accurate tracking corrections. By referencing the image of a star near your subject to the reticle in a suitable eyepiece, you can verify that your mount is tracking properly and get far better results than you could otherwise. Ideally, the focal length of your guide scope should be at least as long as that of the optics you are using on your camera. In most cases, it is really preferable if the guide scope is several times the focal length of your camera lens. This will allow you to more accurately see and correct tracking errors, resulting in more detailed photos. If you mount your camera "piggyback" on your main telescope, your telescope can then be your guide scope.

The most common tracking errors occur in the right ascension (RA) axis. Most of these errors are associated with imperfections in one or more of the drive's mechanical components. The drive components typically rotate during normal use, so it is common for the most obvious tracking errors to recur with each full rotation of an imperfect component. This type of error is called "periodic error" due to its periodic nature. Some newer telescopes have circuits which "remember" an astrophotographer's guiding corrections and compensate for most periodic error, but even this is not enough to completely eliminate the need for guiding, particularly when long focal length optics are used for photography.

If you are shooting through a normal or wide angle lens, adequate tracking corrections can usually be made with good mechanical slow motion controls; however, if you use a relatively long telephoto lens, your tracking corrections will have to be more precise. Here, a manual tangent arm control having a lot of mechanical reduction can be useful, but the best solution is an electronic drive rate corrector for the mount's motor drive. A drive rate corrector will also be useful when or if you get into more advanced astrophotography such as guided deep sky photography through your telescope.

A drive rate corrector (often just called a "drive corrector") facilitates correction of tracking errors by varying the right ascension (R.A.) drive motor speed. This is typically done with an electronic circuit that is either in a separate box or built into the equatorial mount itself. With a proper drive corrector, you can make tracking corrections in at least one axis without physically touching your telescope or its mount.

If you do not have a drive corrector, you can make tracking corrections by interrupting your exposure, moving the mount manually in right ascension, then resuming your exposure. If you use a telephoto lens, it is not unusual to have to interrupt your exposure as you make manual tracking corrections. This exposure interruption is often necessary because the mount may vibrate excessively when you touch it. In addition, most manual slow motion controls are too coarse to allow you to reliably make a proper fine correction on the first try. By interrupting your exposure, you can take as long as you like to make a manual correction. This can be tedious, but it will at least provide a way for you to get results with equipment you may already have.

Corrections in declination may also be necessary, but typically not as often as in right ascension. A motorized declination axis will allow you to move your mount at least a moderate angle in declination without having to touch it. Such a feature is necessary if you are using an autoguider. It can also be handy for other applications such as scanning the lunar surface with a video camera and telescope. Declination motors are convenient, but not always necessary for manually guided film photography. I usually don't go to the expense of getting a declination motor for a given mount if I can easily reach its manual declination control while I am looking in the guiding eyepiece.

It should not be necessary to make declination corrections all that often. In fact, it is seldom necessary to correct errors in declination when an equatorial mount is correctly polar aligned, though there are cases in which atmospheric refraction near the eastern or western horizon can be significant enough to require some correction. I rarely spend more than 5 minutes polar aligning my telescope, yet I have occasionally been able to guide as long as 90 minutes without making a single declination correction.

You need not be this accurately polar aligned to get a good picture, but you may begin to see some field rotation (where the image at the focal surface rotates around your guide star) if you have to make more than about a dozen modest declination corrections in the same direction during your exposure. Taking the time to accurately polar align your telescope will usually make guiding easier and provide a better end result.

Before getting down to the nuts and bolts of aligning an equatorial mount and using it for astrophotography, a few general precautions are in order.

* Never force any controls on your telescope or its mount.
* Never force anything into a housing which contains optics.
* Never point any optical system at the sun unless it has proper front filtration.
* Be sure you have a good grip on your telescope when you loosen the controls, clamps, or attachment screws on either it or your mount.
* Be sure the eyepiece is securely locked in place before you position a telescope.
* Always cap your optics when they are not in use.
* Be sure that a telescope (particularly a computer controlled one) and its accessories or cables won't collide or get caught when the telescope is positioned.

7.1 Polar Alignment

As covered in Chapter 2, an equatorial mount effectively cancels out the earth's rotation by rotating at the same rate, but in the opposite direction. In order to be effective, the polar axis of the mount must be pointed toward the celestial pole so it will be parallel with the earth's axis of rotation. If it is not, it won't accurately track in all parts of the sky. Fortunately, polar alignment is not difficult if approached in the right way.

An equatorial mount may be polar aligned in a number of ways, but only a few methods need to be covered here because the goal of all alignment methods is to get the polar axis of your telescope parallel with the earth's axis of rotation. Some equatorial mounts include dedicated polar alignment telescopes. Such a scope can save time, but accurate polar alignment can also be accomplished without one.

7.1.1 Before You Polar Align, Align Your Telescope to its Mount!

If your mount does not have a polar alignment scope, you will need to point your main telescope toward the celestial pole (plus or minus 90 degrees declination, depending on which hemisphere you are in) during your polar alignment procedure. In this case, one of the first things you need to do is verify that the optical axis of your telescope can be positioned reasonably close to parallel with the polar axis of your equatorial mount. This is something that one should be able to take for granted, but alas, very few commercial telescopes are capable of this without at least some tweaking.

It is very important to properly correct or compensate for any misalignment between a telescope (or polar alignment scope) and mount as soon as you can. Failure to do so can keep you from attaining accurate polar alignment and lead to constant frustration when you take tracked astrophotos.

The alignment between your telescope and mount need not be exact, but it should be within $\frac{1}{4}$ of a degree or so. This will allow you to use an eyepiece which provides up to about 80× magnification when you align, yet still see a point parallel with your mount's axis in the field when your telescope is pointed to 90 degrees declination (or –90 degrees if you are in the southern hemisphere).

If your telescope has to be modified in order to be properly aligned with its mount, the good news is that such a modification usually has to be made only once for a given telescope. Implementing this alignment if needed (or adapting a polar alignment scope to your mount) is one of the most important improvements you can make to a photographic telescope mount. If you (or a machinist) first make proper modifications or otherwise compensate for misalignment in a telescope mount, polar aligning it will usually be substantially faster and easier than it would be otherwise. It will also tend to improve polar alignment accuracy, making it far easier to guide your photos.

Before you start, it is helpful to make sure that your telescope is properly collimated, and to carefully collimate it if needed, provided it has proper provision for this adjustment. This is particularly recommended prior to aligning a reflector telescope to its mount because collimation of a reflector actually makes small changes in where the optics are pointing in relation to their tube!

When a good small to medium size telescope is properly collimated, the high magnification image of a fourth to seventh magnitude star during good atmospheric seeing conditions should look similar to (or at least reminiscent of) the Airy disk (the brightest blob of light) and diffraction rings shown in Chapter 4. If the telescope is not properly collimated, a focused star image will usually appear to be asymmetrical, with most or all of the extraneous light on one side of the Airy disk. If the collimation error is only minor, the diffraction rings will completely surround the Airy disk but they may just appear heavier on one side.

In evaluating collimation, remember that most real world amateur telescopes (particularly larger ones) don't have optics good enough to produce a really good Airy disk and diffraction pattern. Those that do may often be hindered by atmospheric conditions or tube currents. If you are in doubt about whether your telescope needs collimation, just leave its collimation adjustments alone and instead proceed with verifying how well the telescope is aligned with its mount. You can always redo the alignment if you should later collimate your telescope. Most prime focus photography through moderate and slow f/ratio telescopes is tolerant of minor errors in collimation, but long focal length planetary imaging is a bit more picky.

Misalignment between a telescope and mount can arise from a number of imperfections. Some are easy to correct, while others may be more difficult. If you can substantially correct the problem by moving your telescope in declination, then the problem is simply a poorly adjusted declination setting circle.

If you can't fix the problem by adjusting your setting circle or its pointer and the error is off to one side of the declination travel, the cause may be mismatched tube rings or dovetail plates, poor perpendicularity between the right ascension and declination axes, or even misalignment of the telescope optics within the telescope tube itself. The latter two problems are common with some popular fork mounted Schmidt–Cassegrain telescopes.

Whatever the problem, it should be fixed or otherwise compensated for. Sometimes the problem can be corrected by having a qualified person remove a fork arm and file out its attachment holes until they form ovals which can serve to effectively shorten or lengthen it

when it is reattached. Remedies for German equatorial mounts can include adding shims between the tube rings and the mount.

In order for your telescope to be properly polar aligned without a lot of fuss, a star image should appear to rotate about a center that is somewhere within the eyepiece field when the telescope is: (1) used at about 80× magnification; (2) positioned at a setting of exactly 90 degrees declination (or –90 degrees if you are in the southern hemisphere); and (3) rapidly moved several dozen degrees in right ascension.

A good target to use for this test is Polaris or some other star near the celestial pole, since such a star will not move much with time. Ideally, the image will appear to rotate about the center of your eyepiece field as you move the telescope mount in R.A. If it does not, you should try to get the center of rotation as close to the center of the eyepiece as possible by ever so slightly moving your telescope in declination. If you can get the center of rotation somewhere in the eyepiece field, you will be able to correct for the remaining error during your polar alignment procedures.

When you get the center of rotation as close to the center of your eyepiece as possible, adjust your declination setting circle to read exactly 90 degrees. This is usually a one time adjustment, though there are exceptions (such as collimating your telescope) which may bring about the need for you to reset it. If your mount's declination circle is not adjustable, make careful note of the setting it reads for future reference. You may instead want to add an appropriate mark to the circle or its pointer. Such a mark for 90 degrees declination will then be the point of reference from which you set the declination axis for the coordinates of Polaris every time you polar align your mount.

With the declination still at the above 90-degree setting, manully rotate the mount back and forth through several dozen degrees of right ascension while looking in the eyepiece and note where the apparent center of rotation is in the field of view. When you actually polar align your telescope, you will be compensating for your mount's errors by centering Polaris (or whatever star you choose to align on) in this part of your eyepiece field rather than in the middle. Carefully noting this position in the eyepiece field is the last of the one time stuff you will usually have to do with a given mount.

You can improve the accuracy of your subsequent polar alignment if you add a reticle (an off center one if necessary) to your eyepiece which indicates the area you will use to "center" Polaris. Another way to improve accuracy is to instead use an eyepiece having a field of view in which the center of apparent field rotation falls right on the edge of the field of view.

The reference point for Polaris will usually change in the eyepiece if you rotate your star diagonal or telescope tube. Therefore, it is best to rotate your star diagonal attachment or telescope tube to a standard position from which you can comfortably look into your eyepiece each time you start your polar alignment procedure. Then, just as mentioned above, rotate the mount through several dozen degrees of right ascension while looking in the eyepiece, note where the center of rotation is in the eyepiece field, and use that point as the "center" when you position the image of Polaris in your eyepiece.

If your telescope has a dedicated polar alignment scope, it may not hurt to check out how accurately it is aligned with your mount. You can do this by comparing objects in the field of your alignment scope with those in the apparent center of rotation of the field of the eyepiece on your main telescope when you rotate it in right ascension in the way mentioned above. If you are using a polar alignment or finder scope to polar align something other than a conventional equatorial mount (such as a barn door mount), you can use the same method to verify that the alignment scope is parallel with the polar axis.

Another way to check out a polar alignment scope is to use it to polar align your mount, then use the "drift method" covered below to see how well the mount is polar aligned. Fortunately, dedicated polar alignment scopes which fit inside the polar axis are seldom seriously misaligned because there are fewer things to go wrong in manufacture. Minor misalignment can usually be compensated for by just using a different reference point on the scope's reticle. A *finder scope* having a built-in polar alignment reticle can be another story, since such a scope is potentially subject to even more misalignment problems than the main telescope.

7.1.2 Polar Aligning Your Aligned Mount

If your equatorial mount has a polar alignment scope, its instructions will usually cover how to use it to achieve reasonably accurate polar alignment. In the absence of instructions, you can typically get acceptable alignment in the northern hemisphere by positioning Polaris in a non-inverting alignment scope so it is on the side of the scope's reticle which is *away from* the brightest star in the bowl of the Little Dipper (Ursa Minor). If your polar alignment scope inverts the image, you will instead want to position Polaris on the side of the reticle which appears to be *toward* the brightest star in the bowl of the Little Dipper.

The rest of this section will assume that your equatorial mount does not have a polar alignment scope, or that you want to check the scope's accuracy by using alignment methods which utilize the main telescope. For clarity, whenever the term "center" is used in association with positioning Polaris in an eyepiece field, it always refers to the part of the field around which the image appears to rotate when the telescope mount is pointed to 90 degrees declination and rapidly moved in right ascension.

7.1.2.1 Polar Alignment by Stellar Coordinates

Most popular polar alignment methods involve setting your mount to the coordinates of Polaris, then moving the base of your mount (but not changing the right ascension or declination settings) to get Polaris in the center of your field of view. A prerequisite to doing this usually involves setting your right ascension setting circle to agree with current star positions. This can be done in a couple of ways. One is to first point your mount's polar axis in the general direction of Polaris (within a degree or so is usually good enough), then point your telescope to a star within a couple of dozen degrees of the celestial equator and set your right ascension circle to agree with the star's coordinates.

Once your right ascension circle is properly set, all you have to do is set the telescope mount to the coordinates of Polaris, then move its base (without changing the mount's right ascension or declination settings) until Polaris is in the "center" of the eyepiece field. This will usually get your mount polar aligned to an accuracy of better than about 1/20 of a degree; accurate enough for most astrophotography. For reference, epoch 2000 coordinates for Polaris are 2 hours, 31.5 minutes RA; 89 degrees, 16 minutes declination. Repeating the procedure from the point where you adjust your right ascension setting circle can improve alignment accuracy even more.

The actual position of Polaris changes a little from year to year due to precession in the earth's rotation. In everyday terms, precession is a slow wobbling like you would see in a toy top as it is spinning. The earth's precession is very slow, causing its polar axis to describe a 47 degree diameter circle (twice the 23.5 degree tilt of its axis) in relation to the stars over a period of about 26,000 years. This gradually changes which stars are near the north celestial pole, but the change is slow enough that amateur astronomers can easily make do with star charts which are updated only once every few dozen years.

Due to its proximity to the celestial pole, the right ascension of Polaris can change significantly over a person's lifetime even though its linear displacement from precession is only a fraction of a degree. As shown in Fig. 7.2, the right ascension coordinates of Polaris will change by about two hours (30 degrees) between 1950 and 2050! By 2050, the coordinates of Polaris will be 3 hours, 49 minutes RA and 89 degrees, 27 minutes declination.

By way of review, your declination setting circle should have previously been set or marked to agree with where your telescope is actually pointing. When setting the coordinates of Polaris, the true position for 90 degrees declination should be used as the offset reference point when you move your mount to the slightly lower declination setting of Polaris. In addition, the point in your eyepiece field in which Polaris is "centered" should be the point around which the field appeared to rotate when you previously checked the alignment of your telescope with its mount.

Polar alignment is not difficult, but it has a reputation for being a real pain, partly because many people try to do it without first checking for misalignment between the telescope and its mount. Once any misalignment problem is properly compensated for, accurate polar alignment can become very easy. Also, it is not usually necessary to spend a lot of time trying to polar align your telescope down to a gnat's eyelash. Aligning it to an accuracy of 1/10 to 1/20 degree (about 1/10 of the field of view at 80×) is typically adequate for all but the most demanding applications.

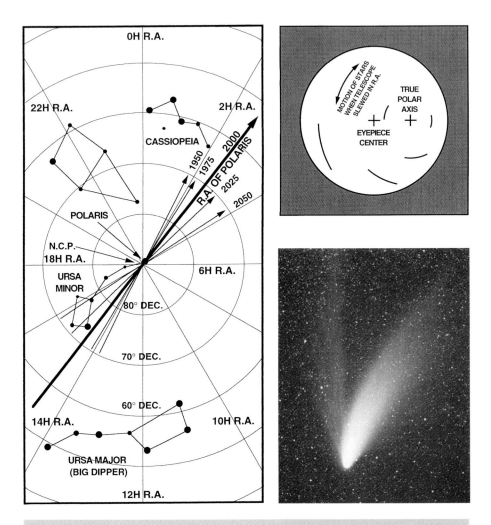

Figure 7.2. Left: This chart shows that stars such as those at one end of the "W" in Cassiopeia, the end of the handle in the "Big Dipper" (Ursa Major), and the bowl of the "Little Dipper" (Ursa Minor) are handily references for quick polar alignment because their right ascension is either nearly the same as, or opposite from (180 degrees from), that of Polaris. Epoch 2000 stellar coordinates and the right ascension of Polaris for the years 1950, 1975, 2025, and 2050 were provided by Carina's Software's Voyager II program. Upper right: The point at which you position Polaris in your eyepiece during polar alignment should be the point around which the field appears to rotate when the telescope mount is positioned at 90 degrees declination and rapidly moved in right ascension. This is true even if it means that you don't use the center of the eyepiece field. This drawing illustrates the situation for a moderate degree of misalignment between a telescope's optics and its mount's polar axis. Lower right: A telephoto lens of modest focal length can provide stunning images of bright comets and other subjects, but really good results depend on accurate polar alignment. This cropped image is from a 10 minute exposure on hypered Fuji 400 film. The 180 mm f/2.8 lens used by Richard Payne to shoot this detailed image was riding piggyback on a properly aligned equatorial mount. Photo courtesy of Richard Payne.

A less accurate but still effective method of polar alignment in the northern hemisphere is to just position your telescope so its polar axis points in the general direction of Polaris, then move the telescope in right ascension until its declination axis is reasonably close to perpendicular with a plane which passes through your telescope, Polaris, and either the end star in the handle of the Big Dipper or the eastern star (in terms of right ascension coordinates) in the "W" of Cassiopeia. Right ascension coordinates for the reference star in Cassiopeia is only a few degrees from those for Polaris. Another star you can use is the brightest star in the bowl of the "Little Dipper" (Ursa Minor). It is actually a better reference than the previous stars because precession will soon cause it to be on exactly the opposite side of the celestial pole from Polaris.

After you position your telescope in right ascension so its motion in declination roughly sweeps across a line between Polaris and one of these reference stars, you can just set the mount to the declination of Polaris and position its base (without again moving the mount in RA or declination) until Polaris is "centered" in the eyepiece. Repeating these last parts of the procedure can provide very accurate polar alignment because you start out with the polar axis closer to the correct position the second time through. Offsetting from the right ascension of reference stars to exactly that of Polaris can provide even more accuracy.

This fast and easy polar alignment method is what I use for my own astrophotography. It works because a line passing through the reference stars and Polaris nearly intersects the north celestial pole (N.C.P.).

It is useful to remember that Polaris is on the side of the celestial pole toward Cassiopeia, so if Cassiopeia is above Polaris and Ursa Major is below it, you would move your telescope predominantly upward (in relation to the northern horizon) from 90 degrees declination to the declination setting for Polaris. If instead Ursa Major is higher than Cassiopeia, you would move the telescope down a little from 90 degrees declination (away from Ursa Major) when you set it to the coordinates of Polaris.

If you repeatedly set up your telescope at the same latitude, either of the above methods can be used in conjunction with a level on your mounting base. In this case, you can level your mount, set the coordinates of Polaris, then just move your mount in azimuth until Polaris is in the "center" of your eyepiece field.

Other polar alignment methods include one in which you reference to relatively dim stars which are even closer to the celestial pole than Polaris, but this method is not always desirable because light pollution or local thin clouds can sometimes keep you from seeing these dimmer stars.

7.1.2.2 *The Drift Method*

The "drift method" is a popular way to verify accurate polar alignment, but it can also be used to actually polar align a telescope mount at times or locations in which one cannot see the celestial pole due to localized clouds or terrestrial obstructions. If you use the drift method, you should first polar align your telescope as accurately as you can without a lot of fuss.

The drift method works because poor polar alignment results in tracking errors which are greatest in zones of the sky which are perpendicular to the direction of greatest misalignment between the equatorial mount's polar axis and the earth's axis of rotation. This corresponds to two opposing points on the celestial equator, one of which is usually above the horizon at any given time. At these points, the "equator" of the equatorial mount's polar axis intersects the celestial equator. Here, the tracking of a misaligned mount will diverge from the motion of the stars, causing a marked drift in declination. This drift is what allows you to check the accuracy of your mount's polar alignment.

With the drift method, what matters is the drift in declination. If the star image moves in right ascension, just recenter the star in that axis while leaving the declination control alone. It is worth noting that the star image should not move too far in right ascension if your mount is even close to being polar aligned. If the image rapidly moves clear out one side of the eyepiece field, you should check to see that your drive is running and that it is set on the correct drive rate (some drives have a lunar rate setting which differs substantially from the sidereal rate) and for the correct hemisphere.

The amount of declination drift you are looking for is rather small; perhaps only a small fraction of a millimeter in 5 minutes if you are using a 2000 mm focal length telescope; even less drift if you are using a shorter focal length instrument. Since you are looking for a small amount of drift, it is best to use a reticle eyepiece which will provide a magnification of 50× or more. If the drift is so fast that you can easily detect it in less than a minute, the polar alignment of your mount is way off and it may take a few tries at repositioning it before you get aligned properly. Unless noted otherwise, the directions of drift shown below are for observers in the northern hemisphere.

To start out, point your roughly aligned telescope at a star near the intersection of the meridian and celestial equator, then center the star image in a reticle eyepiece on your main telescope. If necessary, verify which directions in your eyepiece field correspond to a north and south direction of drift and recenter the star when you are finished. Now, just sit back and wait, or, walk around and chat with fellow astronomers for a few minutes; then look in the eyepiece again to see if the star is still centered on the reticle.

> If the star near the intersection of the meridian and celestial equator appears to drift south, the telescope mount's polar axis is pointing east of the celestial pole. If the star instead appears to drift north, the mount's polar axis is pointing west of the celestial pole. If you noticed no drift, your mount is probably aligned from east to west. (For observers in the southern hemisphere: If the star appears to drift north, the mount's polar axis points east of the *southern* celestial pole since the opposite end of the polar axis obviously has to be misaligned in the opposite direction.)

Next, point your telescope at a star close to the celestial equator and about 20 degrees above the eastern to east-southeastern horizon. (A star well above the horizon is used in order to minimize the effects atmospheric refraction may have on the drift.) Center the star image in your reticle eyepiece, wait a few minutes, then look to see if it is still centered.

> If the star appears to drift south, the telescope mount's polar axis is pointing below the celestial pole.
> If the star appears to drift north, the mount's polar axis is pointing above the celestial pole.

If the star image moves only a small distance (less than about 1/10 mm on the reticle in 5 minutes at a focal length of 2000 mm or more) when the telescope was at either position, your mount is aligned accurately enough for serious astrophotography. If the star image moved a lot more than that, you may want to align your mount more accurately. To do so, reposition the mount only slightly (usually only a fraction of a degree) in the appropriate direction.

If you think your mount is accurately polar aligned and you are using only the drift method to verify its alignment, it can't hurt to just start taking a picture and keeping tabs on how often you have to make a declination correction. If you don't have to make very many corrections, your mount is probably aligned pretty well and you will have saved a lot of time. A frame of film is cheap when compared to the value of your time.

7.2 Piggyback Astrophotography

The sky is full of constellations and a number of interesting deep sky objects which are larger than the field covered by most telescopes. The latter objects are relatively dim, so long exposure times are typically required to get a good image. Piggyback astrophotography is one of the easiest ways to get pictures of deep sky objects, partly because camera lenses tend to have faster f/ratios than most telescopes. A fast f/ratio in turn permits the use of a shorter exposure time. Piggyback photography is also useful for imaging planetary conjunctions and other events which occupy a relatively large part of the sky.

"Piggyback" is a popular description for this type of photography because your camera is most often "piggybacked" right on your telescope. This allows you to use the telescope for guiding during your exposure. The term piggyback can also apply to other configurations, including setups in which the camera and telescope are mounted side by side on a plate; where a camera is mounted on the counterweight shaft of a German equatorial mount; or, where a large camera lens or Schmidt camera is used in place of the main telescope and a smaller guide telescope actually rides piggyback on it. Using a separate telescope to guide while taking pictures through your main telescope is not usually called piggyback photography even though it is similar in concept to the latter setup.

It is not unusual for piggyback photography to be used in taking pictures which cover anywhere from the entire sky to the relatively narrow three or four degree field of a long telephoto camera lens. With so many degrees of freedom in regard to lenses, exposure times, and films, it is little wonder that piggyback photography is one of the most popular forms of astrophotography. The whole sky is before you. All you have to do is go out and start taking pictures.

With a given lens and film, your results will vary according to the exposure time you use. If you use an f/ratio of 2.8 and a film speed of ISO 200, an exposure of about a minute through a normal to moderately wide angle lens will usually image the major stars in a constellation. An exposure of 10 minutes will usually provide an image which is reminiscent of what the same area looks like with the naked eye. A 30 minute or longer exposure from a dark site will image many features which are too dim to easily see with the unaided eye.

Exposures of several tens of minutes through relatively long focal length camera lenses can reveal objects and features similar to those you would see through a typical pair of binoculars. Longer exposures will usually reveal more features. A dark site is important if you want to photograph dim objects. If you do not take your pictures from a really dark site or use a light pollution rejection filter at a moderately dark site, your images will not reveal much in the way of dimmer deep sky objects.

7.2.1 Positioning Your Camera

An important consideration for piggyback photography is the stability of the piggyback mount itself, since excessive flexure between the telescope and camera can result in poorly tracked images. Another consideration is the possibility that an optical component in your telescope can move even when its tube does not. If you are using an internally focusing telescope to guide, movement of an internal optical component (such as the primary mirror in an internally focusing commercial Cassegrain telescope) during your exposure can cause visible errors in piggyback photos taken with telephoto lenses, though it is not usually a problem for wide angle shots. If an internally focusing telescope has provision to lock down the component which moves during focusing, it is a good idea to secure it before you use it to guide long focal length piggyback photos. By eliminating flexure and mirror motion, you can avoid many unpleasant surprises in your photos.

a

b

c

Figure 7.3. Different camera lenses can capture different aspects of the night sky. **a** A 20 mm wide angle lens covers a vast expanse of the sky and can make our own summer Milky Way look almost like a distant galaxy looming above the horizon. This 30 minute exposure at f/5.6 was shot on ISO 1600 print film. **b** A 10 minute exposure with a 35 mm f/2 lens on the same film provides enough image scale to pick out a few individual nebulae. **c** An inexpensive 400 mm lens captures a close up view of the Lagoon nebula (M8) and Trifid Nebula (M20). The exposure was 50 minutes at f/6.3 on Ektachrome 400 film. A higher quality lens will typically produce less bloating of the bright foreground stars.

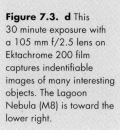

Figure 7.3. d This 30 minute exposure with a 105 mm f/2.5 lens on Ektachrome 200 film captures indentifiable images of many interesting objects. The Lagoon Nebula (M8) is toward the lower right.

d

When guiding your photo, it is usually best to guide on a star which is in or near the area covered in your piggyback picture. This will minimize the effects field rotation (from inadequate polar alignment) can have on your image. Field rotation is concentric with your guide star, so a star close to your subject will provide a shorter radius from your subject the center of any field rotation. The shorter radius will in turn result in shorter streaks during a given exposure time. Of course, the best policy is to just make sure your mount is accurately polar aligned, since this can eliminate field rotation altogether.

A stable ball head is a good choice for positioning a camera with a wide angle or modest telephoto lens. This type of head makes it easy to point your camera in a desired direction as well as facilitating rotary orientation of the camera for composition. Some ball heads having good stability are made by Bogen, Canon, Gitzo, and Slik. A good tripod head of conventional design will usually provide just as much stability, but positioning your camera the way you want may take longer.

If you are using a wide angle lens, you should look in the camera viewfinder while shining a light on your telescope in order to verify that part of the telescope or its mount are not included in your picture. Longitudinal adjustment in a piggyback mount can be very useful for positioning a wide angle camera relatively close to the front of a telescope where it will not be obstructed. If you have a fork mounted telescope, you will typically have to add counterweights on the side of the tube which is opposite your camera.

Counterbalancing your camera can have a lot to do with how well your mount tracks. In the right ascension axis, your telescope and its piggybacked camera should be balanced pretty well, but not quite exactly. If you have a little of the weight offset in a way that will gently pull the assembly toward the east (in RA), it can help take up backlash in your drive. In the declination axis, you can get away without balancing things quite as accurately. Another precaution is not to start your exposure until at least 2 or 3 minutes after you position your telescope. This will allow some time for the drive to take up any loose backlash in its drive train, typically resulting in more reliable tracking.

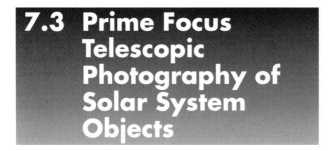

7.3 Prime Focus Telescopic Photography of Solar System Objects

The long focal length of most amateur telescopes provides enough image scale to get a good photograph of the full disk of the sun or moon. Planetary images at the prime or Cassegrain focus of an amateur telescope are still a bit small, but there are times when a small planetary image is adequate. Examples are conjunctions and occultations such as those pictured in Chapters 2 and 9. A larger planetary image can be obtained with auxiliary optics including teleconverters, Barlow lenses, or by using an eyepiece for projection or afocal photography. Whether or not you have any of these auxiliary optics, there are plenty of objects out there to photograph.

Some objects can be photographed during the day. The sun is a surprisingly dynamic subject which can be observed and photographed with appropriate front mounted filters. Most small and moderate size amateur telescopes provide a solar image which is small enough to fit the 24 × 36 mm film format of a full frame 35 mm camera. Long focal length instruments may provide too large an image for the 35 mm format. In this case, a tele-compressor can usually be used to provide a smaller solar image if the telescope has enough back focus.

The moon and some planets can also be observed through full daylight. With a telescope, you can see planets including Mercury, Venus, Mars, and Jupiter. Saturn can sometimes be detected in a telescope during the day, but observing any detail on it is another matter. Brighter stars are also visible through a telescope on a clear day, but they are less appealing photographic subjects than planets.

Daytime photography of the moon and most planets does not provide as much image contrast as a photo taken at night, but there are times when the position of a planet or the timing of an astronomical event require daytime photography. Examples include photographing an occultation that occurs during the day, or shooting Mercury or Venus when their elongation from the sun is minimal. Venus in particular is so bright that you can easily get good pictures of it during the day. Visual observation of Venus is also easy during the day; perhaps even easier than at night because the brightness of the daytime sky can help mask the effects of glare and some optical aberrations.

Nightfall brings the opportunity to observe and photograph different attributes of the moon and planets, plus a host of other objects which are too dim to see during the day.

The moon is one of the most photographed celestial objects. Its appearance gradually changes from a broken crescent so thin it can hardly be seen, to a bright full moon which dominates the night sky. In between these extremes, the moon can have a variety of appearances; from a thicker crescent to a half illuminated moon having a terminator rich with detail, to a commanding gibbous phase.

When observed at night, a crescent moon appears to cradle the rest of its surface, which is dimly illuminated by sunlight that is reflected from the earth. This dimmer "earthshine" is brightest when the moon is a crescent because the earth's phase (as observed from

the moon) is the inverse of the moon's phase as seen from earth. Earthshine can also be observed and photographed on the moon during its half and moderately gibbous phases, but it is more difficult to detect.

As the moon makes its appointed rounds during its orbit, it is occasionally eclipsed as it passes through the earth's shadow. At other times, the moon passes near planets and brighter stars. On rare occasions, it will cover a planet, a relatively bright star, or even the sun.

A good telescope or sufficiently long camera lens having an aperture as small as 4 or 5 cm can provide good lunar phase photos. Pictures of the moon can also record the effects of lunar libration, which is a slow wobble of the moon which is induced by its elliptical orbit. This is evident in the montage photo of different lunar phases. Some of the photos were taken during different lunar cycles. Libration can be detected as differences in the distance between the lunar "maria" (or seas) and the lunar limb on the right side of each image.

Figure 7.4. Even a small telescope can provide good images lunar phases if it is properly mounted and vibration from the camera shutter is minimized. This montage shows a few phases of the waxing moon; thin crescent; thicker crescents; half illuminated; three gibbous phases; and full. The thin crescent of a one day old moon at left is a $\frac{1}{4}$ second exposure on Ektachrome 200 film through a 9 cm f/11 Maksutov-Cassegrain telescope. The rest were shot on the same type of film through a 10.2 cm f/15 refractor, using the exposure values recommended in the astrophotography exposure guide in Appendix A of this book.

7.4 Photographing Specific Solar System Objects Close Up

The sun, moon, and planets are remarkable photographic subjects, but capturing their fine detail typically requires a longer focal length than the original focal length of most amateur telescopes. Planets have such a small angular size that they are imaged as little more than a dot at the prime or Cassegrain (i.e. direct) focus of a small to moderate size amateur telescope. Fortunately, a vast majority of telescopes are compatible with auxiliary optics which increase their effective focal length, resulting in a larger image on the film. This in turn provides a better end result because less enlargement of the film image is required.

Some solar system objects also have features which are best seen and photographed through specific types of filters. Here again, most amateur astronomical telescopes are compatible with available filters. With the exception of solar filters, most filters are used behind the telescope. This allows them to have a relatively small size in relation to the telescope aperture, which in turn reduces their cost. Color filters used for planetary observation and photography can be even smaller since they are typically used on the eyepiece. Afocal photography can simplify the attachment of simple filters, since you can use either those which fit the telescope eyepiece or the type which screws onto the camera lens. If you do not have either type of filter, you may be able to use colored acetate between the camera lens and your eyepiece.

7.4.1 The Sun

A safe front mounted solar filter will reveal that the sun's surface is often host to sunspots which are comparatively dark compared to the rest of the solar surface. Sunspots appear and disappear over a period of days or weeks, yet no two are exactly alike. Most last long enough that you can watch them gradually transit the solar disk from day to day as the sun rotates. Sunspots can occasionally be relatively large and plentiful, with the largest ones being large enough to see even if you just look directly through a solar filter (you *must* use a proper solar filter) with your otherwise unaided eye. The frequency and size of sunspots tends to increase and decrease during a cycle of about eleven years.

A photo of the entire solar disk will show that sunspots are present, but additional focal length (and the larger resulting image scale) will provide detail of the dark umbra and lighter penumbra in most sunspots. (In this context, the terms umbra and penumbra refer to features in the sunspots themselves rather than to umbral and penumbral shadows, as in the very different case of an eclipse.) Using a Barlow lens, eyepiece projection, or the afocal method will allow you to get larger sunspot images.

Photographs taken through a telescope which is used in conjunction with appropriate auxiliary optics and filtration will reveal that the sun's surface has a granular texture in which each granular feature is only a couple of arc seconds across. This structure is so small

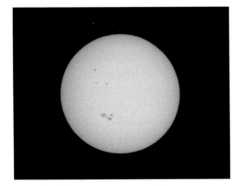

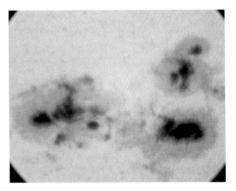

that a very long focal length (typically at least 10,000 mm) is required to clearly image it on film.

Lower contrast solar features include flage, a subtle feature which differs only a little from the brightness of the surrounding solar surface. Flage is usually easiest to see near the edge (limb) of the sun.

Some people shoot long focal length solar photos without any front solar filter. This comparatively risky technique is usually done in conjunction with either a rear projection screen or a rear filter. The method can be acceptable if you don't ever look into either the telescope or your camera's viewfinder at any time other than when a safe visual solar filter is securely attached to the front of your telescope. If you use this method, you should use a safe visual solar filter to compose your picture, then briefly remove it only while you shoot the photo.

I used this method to get a fast shutter speed for the afocal sunspot close up in this section. The camera and lens were positioned on a separate tripod behind the telescope and its eyepiece. To acquire the subject and compose the picture, a safe front solar filter having a transmission of about 1/100,000 was used to reduce the image brightness to a safe level. To shoot the picture, two polarizing filters were placed between the telescope eyepiece and the front of the camera lens. Then, I removed the front solar filter and took the picture without looking in the camera. The front solar filter was replaced immediately after the exposure. The relatively fast 1/125 second shutter speed minimized cumulative atmospheric seeing effects on the image.

When using this technique, it is important to minimize the amount of time direct sunlight can enter your telescope. Too much direct sunlight can introduce tube currents. In addition, if your telescope, camera or eyepiece have plastic light baffles, the concentration of unfiltered sunlight by your objective could melt them!

Figure 7.5. Left: The 1420 mm focal length of an 8.9 cm f/16 Maksutov-Cassegrain telescope provides a large enough solar image to clearly show this exceptionally large group of sunspots. Right: The afocal method was used with the same telescope to get this close up of the sunspot group at an effective focal length of about 25,000 mm.

A relatively weak front solar filter can facilitate close up solar images with less effort. Such a filter is not safe for visual observation, but using one for photography is acceptable if you don't ever look into either the telescope or your camera's viewfinder when the safe visual solar filter is removed. To compose your picture, you should use a safe visual solar filter in front of the telescope. When you shoot your picture, you can briefly replace the safe filter with the weaker filter.

Other aspects of the sun can be seen and photographed with more specialized filters. When a good hydrogen alpha filter is used in accordance with its instructions, solar flares, prominences, and other interesting solar features can be observed, photographed, or otherwise recorded even in the absence of an eclipse. Prominences in particular are fascinating and ever changing, though they do not usually move or change fast enough that you can observe their motion in real time. Prominence activity generally increases and decreases along with the sunspot cycle, but it is rare for the sun to have no visible prominences at all. A hydrogen alpha filter having a correct band pass of narrower than an angstrom will reveal an even broader range of dynamic features, some of which can be seen against the solar surface.

Most hydrogen alpha filters have front and rear components: a front red filter which rejects most of the unwanted wavelengths, and a rear component which includes a more precise filter. The rear filter component is relatively sensitive to the incident angle of light to its surface, so most have to be used at slow f/ratios to provide the best images. This is because light converging from the edge of the telescope aperture intersects the filter at a different angle than light from the center of the aperture.

7.4.2 The Moon

The original image from a typical amateur telescope is large enough to provide a good photo of the entire moon, but a larger image is necessary if you want to get detailed pictures of individual lunar features. These features provide a wealth of subjects for the astrophotographer. Craters are the most popular subjects, but other features include mountains, large cliffs or valleys, rills, maria, and streaks such as those radiating from the crater Tycho. Most lunar craters are best photographed when the terminator (the separation between light and dark) is near their location, since the low angle of the local sunlight illuminates them from one side, making their relief appear relatively pronounced. Some of the best detail is visible when the moon is about half illuminated.

For larger images, the effective focal length of a telescope can be increased a moderate degree with a Barlow lens or teleconverter. A significantly greater focal length increase is possible with eyepiece projection. Afocal photography (where you point your camera and lens into an eyepiece) can also provide a larger image. It has the advantage of being compatible with cameras having fixed lenses as well as telescopes having fixed eyepieces.

If you want a smaller or brighter lunar image, you can use a telecompressor if your telescope has enough back focus. A telecompressor will permit a shorter exposure time during a total lunar eclipse or when photographing earthshine on a crescent moon. Methods for calculating focal length with auxiliary optics were covered in Chapter 3.

7.4.3 Mercury and Venus

Planets are very interesting subjects, though they can be more difficult to photograph due to their small angular size. Mercury and Venus go through a full range of phases from a thin crescent to a fully illuminated disk. Little to no surface or atmospheric detail is visible in most amateur telescopes though some observers reposrt occasionally seeing localized variations in brightness on the terminator of Venus. Mercury and Venus are most obvious in the morning or evening (depending on their positions) but they can also be observed through a telescope in broad daylight if you can locate them.

Venus is a naked eye object during the day if you have good eyesight and know where to look. Fortunately, you can usually find it and some other planets during the day through the use of position data obtained from printed works, computer programs, or the Internet; then scan for them with binoculars or use the setting circles on your telescope to dial them in. The orbit of Venus can periodically bring it closer to the earth than any other planet or take it to the opposite side of the sun. Due to its changing distance from earth, the angular size of Venus can vary by more than six times, from about 10 arc seconds at its greatest distance to over an arc minute at its closest distance.

Under the right circumstances, you can get photos or video images of Venus rising or setting. If the horizon is far enough away and the atmospheric conditions are good, you can even capture Venus and terrestrial objects in the same telescopic photograph. To clearly see the phase of Venus in a film image, it is best to use a focal length of at least 2000 mm. An afocal video image of Venus can be acquired with a shorter focal length. In fact, even a modest terrestrial spotting scope may work for video of a crescent Venus.

Long focal length photos which combine Venus and the horizon can be most interesting when Venus is a crescent phase, since this is the time it is closest to the earth and has the largest angular size. A relatively small aperture telescope will give you more depth of field than a large one so the planet and features on the horizon can be imaged together more sharply.

On rare occasions when Venus is within a degree or so of the sun during inferior conjunction, the refracting effects of its atmosphere can cause interesting effects like making its crescent appear to extend well over half

Figure 7.6. Left: It's not the moon setting; it's Venus! Even a small telescope can provide useful planetary images when used in conjunction with the right auxiliary optics. This $\frac{1}{4}$ second exposure on Ektachrome 200 film was taken with a 9 cm f/11 Maksutov-Cassegrain telescope and a 2× Barlow lens with an extension tube which provided a focal length of well over 2000 mm. The relatively small 9 cm aperture was enough to clearly image the phase of Venus, yet also provide enough depth of field to identify trees in the foreground. Right: These images are enlarged to show Venus at an angular image scale of about 3 arc seconds per millimeter. The upper left one was taken through a 9 cm f/11 Maksutov-Cassegrain telescope, the second through with a 20 cm f/10 Schmidt-Cassegrain, the third through a 10.2 cm f/15 refractor, and the last through a 15 cm f/15 Maksutov-Cassegrain. In all cases, a Barlow lens was used to increase the telescope's effective focal length. The respective enlargements from the original photos are 6×, 4×, 2.5×, and 10×.

way around its circumference. Some observers have even reported a complete ring of light at such times. On even more rare occasions, Mercury or Venus may pass directly in front of (or "transit") the sun, appearing to be a dark spot making its way across the solar disk over the period of a few hours.

7.4.4 Mars

The orbit of Mars can bring it comparatively close to earth and take it even farther away than either Mercury or Venus, causing it to have an extraordinarily wide range of angular sizes. Mars also goes through phases, but its thinnest phase is a gibbous one in which it is still well over $\frac{2}{3}$ illuminated. Its rotation reveals different permanent and transient features, depending on when it is observed.

The surface of Mars has been a subject of interest to astronomers for more than a century. Gone are the days when people thought its surface was covered with canals or that Martians were looking back at us, but observing Mars is nonetheless fascinating. It has many interesting features, some of which are easier to see and photograph with various color filters. At other times, you can witness the temporarily obscuration of major features as Mars goes through the throes of global dust storms.

Figure 7.7. The left two images of Mars were taken through an 8.9 cm f/16 Maksutov–Cassegrain telescope. Eyepiece projection provided a focal length of about 14,000 mm and an f/ratio of f/157. Exposure data for the left image was not recorded. The center one was exposed for 1.5 seconds on ISO 100 color negative film. A little more exposure would have been optimum. The right image was taken through the 61 cm refractor at Lowell Observatory in Flagstaff, Arizona on a cold and sparsely attended public viewing night in the 1980s. A Barlow lens was used to provide an effective focal length of about 18 meters. The left two images are enlarged just under 7×. The right one was enlarged about 5× in order to provide the same image scale as the other two. These and all following planetary images in this section are reproduced at an image scale of about 2.5 seconds of arc per millimeter.

7.4.5 Jupiter

Jupiter is a "gas giant" which has interesting belts and festoons that can change radically over a period of time. Its rotation rate is so rapid (less than 10 hours) that its shape is noticeably oblate due to the centrifugal force. Slow changes to its observable gaseous surface include variations in the color of the great red spot and the appearance of larger equatorial belts. Over a decade ago, one of the two large equatorial belts practically disappeared for about a year.

The largest four of Jupiter's many moons are easy to see and photograph (as dots) with modest optics. These moons often appear to transit the surface of Jupiter, and it is usually possible to observe a transiting moon when it is close to the limb of Jupiter with a good 8 cm or larger telescope under good seeing conditions. Observing the entire transit can be more difficult because a moon gets a lot harder to see as it approaches the brighter center of the planet. Fortunately, the shadow of a moon on Jupiter is a lot easier to see.

Figure 7.8. These images of Jupiter show the advantages provided by a large aperture as well as the diminishing returns provided by a large aperture under mediocre seeing conditions. Left: This 6 second afocal exposure on ISO 1600 color print film shows that an 8.9 cm aperture telescope (in this case an f/16 Cassegrain) can provide acceptable planetary images. A 50–250 mm zoom lens was pointed into a 16 mm eyepiece which was used with a Barlow lens on the telescope, providing an effective focal length of 32,000 mm. This image is a 2.5× enlargement of the original. Center: Eyepiece projection with a 32 cm f/6 Newtonian telescope provides substantially better resolution under good seeing conditions. This is a 6× enlargement from the original. Photo courtesy of Pierre Schwaar. Right: This image was taken through the 61 cm refractor at Lowell observatory in Flagstaff, Arizona on a public viewing night and under moderate seeing conditions. A Barlow lens was used to provide a focal length of about 18 meters and this image is about a 4× enlargement. The photo is reasonably good, but it is definitely less than twice as good as the previous photo through a telescope having only half as much aperture. The refractor's chromatic aberration has some detrimental effect, but atmospheric conditions can play a greater role in degrading an image because a large aperture telescope must "look" through a larger column of air. Atmospheric conditions often keep a film image from having as much additional resolution as the proportional increase in aperture would indicate.

Photographing a transiting moon with a small to medium aperture telescope is not particularly easy, but even a small but relatively good telescope can often capture the shadow of a moon on Jupiter's surface. On rare occasions, one of Jupiter's moons will eclipse another. This too can be photographed, but simply observing such an event can be just as much fun.

7.4.6 Saturn

Saturn's beautiful rings are easy to see with the smallest of telescopes, but actually seeing detail in the rings themselves can require more aperture and good seeing conditions. From year to year, the tilt of Saturn's rings as seen from earth gradually changes from edge on to such a pronounced tilt that you can see the rear part of the outer ring over one pole of the planet. The entire cycle takes about 29 years, with two ring plane crossings (or sets of ring plane crossings) occurring in each cycle. The surface of Saturn has a few belts and festoons, though these are not as pronounced as those on Jupiter. A few moons are visible through an amateur

Figure 7.9. The tilt of Saturn's axis allows it to be seen from different relative angles over a period of time. The left image shows Saturn near the time of its 1980 ring plane crossing. The exposure is about $\frac{1}{4}$ second on Ektachrome 200 through a 10.2 cm f/15 refractor and 2× Barlow lens. A short focal length was used because the telescope had no sidereal drive at the time! This image is a 27× enlargement. The "Gaussian blur" filter in Adobe Photoshop was used in two of the image's primary colors to reduce the graininess of my underexposed original image, then the brightness and contrast were increased. The center image is a 3× enlargement of a photo taken through a 20 cm f/10 Schmidt–Cassegrain telescope in 1984. Projection with a 9 mm orthoscopic eyepiece was used to provide a focal length of about 20 meters. The resulting slow f/ratio would have required a one minute exposure on Ektachrome 200, so I used a beamsplitter to monitor tracking. Attenuation by the beamsplitter doubled the required exposure time to two minutes. The right image was taken a few years later, when the rotational axis of Saturn was near its maximum tilt toward earth. The 16× enlargement is from an afocal photo through an 8.9 cm f/16 Maksutov–Cassegrain telescope. Specific exposure data was not recorded. Today's faster and sharper films make it possible to get better planetary photos with the same optics; provided you are in an area having good seeing conditions.

telescope, but they are not as bright as those around Jupiter due in part to Saturn's greater distance from the sun. Titan is the brightest of Saturn's moons.

Saturn is about 10 times farther from the sun than earth, so it is considerably dimmer than the inner planets. Due to the inverse square law, Saturn is the square of its proportional distance from the sun dimmer, or 100 (10 × 10) times dimmer than it would be if it were the same distance from the sun as earth. This means that it requires about 100 times more exposure time than a terrestrial subject for a given f/ratio and film speed. If a relatively large image of Saturn is to be obtained from a small aperture telescope, the exposure could literally run into minutes rather than seconds. If such a long exposure is required, you can use a beamsplitter or suitable guide scope to monitor how well your mount is tracking. If the exposure is not too long, you may be able to get by without any guiding at all, or you can try using an off-axis guider to guide on one of Saturn's moons.

7.4.7 Uranus, Neptune, and Pluto

Uranus and Neptune appear to be indistinct greenish disks in amateur telescopes, so neither one is a particularly popular photographic subject. Pluto is so small that its disk cannot even be resolved with amateur telescopes. It looks just like a dim star in either the eyepiece or a film image; however, two observations or photos made at sufficiently differing times will show it in a different location, revealing that it is not a star.

7.4.8 Asteroids, Comets, and Beyond

Like Pluto, asteroids have such a small angular size that no shape or surface detail can be resolved with an amateur telescope. Unlike Pluto, a typical asteroid orbit is well inside that of Jupiter. Due to the faster orbital motion of an asteroid, it is not uncommon for one to appear streaked in an exposure of moderate duration. Comets can also visibly move during an exposure. The general techniques for photographing

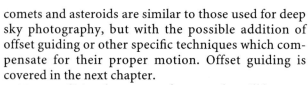

comets and asteroids are similar to those used for deep sky photography, but with the possible addition of offset guiding or other specific techniques which compensate for their proper motion. Offset guiding is covered in the next chapter.

Many realities about astrophotography will become apparent as you shoot your own photos. Your first astrophotos may or may not turn out to be of the best technical quality, but the experience of just shooting them can be rewarding. As you become more experienced, you will no doubt get better images, but let's not forget that for many amateur astronomers, the object is just to have a good time.

Each new picture you get or each new technique you learn is but a step in the journey of amateur astronomy, and the journey itself should be every bit as fun as reaching any goals you may set along the way.

Chapter 8
Guided Astrophotography Through a Telescope

The last chapter showed that piggyback photos can cover relatively wide areas of the sky, but that they don't capture much detail on deep sky objects having a small angular size. By taking pictures through your telescope instead of a camera lens, you can get larger and better images of deep sky objects.

Deep sky astrophotography through your telescope is both one of the most demanding and most rewarding types of astrophotography. This is how you get those interesting shots of individual galaxies, star clusters, and reflective and emission nebulae. These "deep sky" objects are typically identified by an "NGC" (for New General Catalog) number. Some of the brightest and most popular objects are commonly identified by an "M" or "Messier" number. These 100 plus objects are from a catalog the French astronomer Charles Messier published in 1784.

The state of the art in amateur deep sky photography is approaching the level of quality that professional observatories had published not so long ago, but most of this cutting edge amateur work still involves the use of relatively large telescopes or expensive wide field astrographs.

Many small or relatively inexpensive telescopes can also be used for deep sky photography. Pictures taken through them may not fit the description of cutting edge, but they can definitely be good. What matters most is whether or not you will be satisfied with the resulting pictures, based on what you can realistically expect from your equipment. By shooting your own photos, you end up with a final image that is cropped and displayed the way you want. Best of all, you will have captured a piece of the sky – with your own camera!

You can most likely start shooting deep sky photos through your telescope if (1) a camera body can be attached directly to the telescope; (2) the telescope has a sidereal drive; and (3) you have some means of guiding and accurately making appropriate tracking corrections.

Even if you plan to get a better telescope later on, it can still be worth using a telescope you may already have to at least get familiar with the related techniques. If you lack guiding accessories for your present telescope, you may be able to find suitable accessories which can be used with both it and any telescope you may plan to buy or make later. By starting to take pictures now, you can gain experience in deep sky astrophotography. Then, if you get a bigger or better telescope, everything about using it will probably seem easy!

The first group of deep sky photos in this chapter were taken through a telescope having only 8.9 cm of aperture. The Questar Maksutov Cassegrain telescope I used was relatively expensive, but it is important to point out that such an expensive telescope is not necessary for deep sky astrophotography.

As covered in an earlier chapter, photos taken at the prime or Cassegrain focus of two reasonably good telescopes won't reveal the difference in resolution, at least in the center of the picture. The difference will only become obvious when each telescope is used at a long enough focal length (and slow enough f/ratio) that its resolution (rather than the film characteristics) will limit the image quality. In most cases, this usually means that you won't see the difference between a mediocre telescope and a good one (at least in terms of sharpness) unless you use them at an f/ratio approaching f/22 or slower, something you probably won't do for deep sky photography. There may be differences in the contrast or off-axis image quality of various telescopes, but price does not always determine which telescope will be the best in these respects.

The majority of small Cassegrain telescopes have a faster f/ratio and less reduction in off-axis image brightness than the 8.9 cm telescope I used. Many Newtonian reflectors and modern refractors also have a faster f/ratio. A fast f/ratio will result in a smaller image for a given aperture size, but many of the larger "showpiece" deep sky objects can be imaged just fine at a focal length as short as 800 mm or so. The price you pay for using a small aperture telescope is that deep sky exposures must be longer than would be necessary for a large aperture telescope of the same focal length.

Getting results with a small telescope is simply a matter of putting in enough guiding time, though the required time can obviously be quite long. For example, the 8.9 cm Cassegrain telescope I formerly used for some of my deep sky photos worked at a focal length of about 1516 mm and an f/ratio of f/17. In some cases, I had to expose deep sky photos for nearly two hours, and even this was not long enough for some objects. By contrast, a 30 cm Newtonian telescope of the same focal length would work at f/5. To get the same result that my 8.9 cm f/17 telescope provides in two hours, a photo through a 30 cm f/5 telescope would require an exposure time of only about 11 minutes!

The difference in exposure time between a large and small telescope of the same focal length is directly proportional to the relative change in aperture area, excluding the effects of reciprocity failure in the film and factors such as any difference in optical transmission. You can calculate the difference in aperture area, but another way to get relatively close is to just count how many f/stops there are between the f/ratio of each telescope, then modify the exposure time by a factor of two for each f/stop.

It is easy to envision aperture area when one telescope is exactly twice as large as another. A 30 cm telescope obviously has four times the aperture area of a 15 cm telescope. If the focal length is the same for either telescope, the large one will be two f/stops faster than the small one. Therefore, the exposure time will only have to be $\frac{1}{4}$ as long for the large telescope in this example.

If instead the f/ratio is identical on both telescopes, the same exposure time will be required with either one, but the larger telescope will provide a bigger image. These relationships to f/ratio and exposure time apply only to extended objects such as nebulae, comets, planets, the moon, and terrestrial subjects because f/ratio (not aperture diameter) is what determines the optimum exposure time for extended objects.

Aperture is what matters when it comes to the exposure required to image individual resolvable stars of a given magnitude. I say "resolvable" stars because Globular star clusters can usually be treated as extended objects unless your telescope has enough focal length for each star to be resolved in a separate grain unit on your film. In the case of an easily resolved open star cluster, aperture will determine how many stars (more particularly, how bright of stars) you can image in a given time with a given film.

Whatever your exposure time, guiding must be performed accurately for its entire duration; however, this does not mean that you have to be a hermit while taking your photos. I frequently carry on conversations with other astronomers in my group while guiding my own pictures. If talking with other observers does not do much for you, you can always bring along a CD or tape player and some headphones, then listen to your favorite music while you guide.

Guiding a long exposure is not all that bad if your guider has a manual shutter. A manual shutter allows you to take breaks during your exposure, literally turning one long exposure into a few short ones. If the thought of guiding for a total of an hour or more still makes you queasy, you can always use a telecompressor or a small, fast, Newtonian telescope. Either of these alternatives will usually facilitate a shorter exposure.

Most astrophotographers who take deep sky photos through a telescope tend to use a relatively conventional equatorial mount, but a few use alternatives like systems based on the principle of "barn door" mounts, Poncet mounts, or other tracking devices.

Deep sky photography at a long focal length places relatively stringent demands on a telescope mount, since the mount's sidereal drive has to accurately track your subject during exposures which can last anywhere from several minutes to a few hours. Few telescope mounts track accurately enough to allow you to just start such an exposure and walk off. Invariably, some form of tracking correction is required.

In order to make tracking corrections, it is first necessary to have some way to detect tracking errors. One way of doing this is to keep tabs on how much the image of a guide star moves in relation to your film. To monitor and correct tracking, you can use a suitable guiding system which is based on either a separate guide scope or an arrangement which allows you to reference to a star image formed by the same telescope which is being used to take your pictures. The latter arrangement can consist of an off-axis guider, partial aperture guider, beamsplitter, or other system.

Actual guiding can be accomplished visually through a reticle eyepiece (preferably one with an illuminated reticle), or you can use electronic means such as an autoguider. An autoguider will literally let you sleep

Figure 8.1. Upper left: Even a small telescope can provide good images of deep sky objects if you take long enough exposures. This 8.9 cm f/16 Questar telescope was used with the shown Versacorp VersaGuider™ off-axis guider to take all of the deep sky photos in this group. As shown, the telescope provides a photographic focal length of about 1516 mm and an f/ratio of f/17. This is slightly longer than the telescope's advertised focal length because the guider requires a few centimeters more back focus than Questar's standard camera coupling. (Increased effective focal length of an internally focusing Cassegrain due to increased back focus is covered in Chapter 3.) Upper right: A 45 minute exposure was long enough to get a bright image of the Perseus double cluster on off the shelf (i.e. unhypered) ISO 1600 color negative film. This image is a 2× enlargement of nearly the full frame. A special declination range extender bracket was required to permit adequate clearance for the camera when photographing an object this close to the celestial pole from the telescope's small fork mount. Middle right: A 90 minute exposure on hypered ISO 1600 color negative film captured the spiral arms of M51, the whirlpool galaxy. The advantage of a larger scale film image can be appreciated by comparing the star images in this photo with the bloated ones in Fig. 7.1b, which is a photo of the same part of the sky that was shot with a 200 mm lens. This and the remaining images in this group are enlarged three times and cropped to show the subjects of interest. Lower left: A 110 minute exposure on the same film provides a good image of the Trifid nebula. I lucked out on polar alignment when I took this image; only one declination correction was required during the entire exposure! Lower right: A one hour exposure on off the shelf ISO 3200 color negative film was enough to image the M46 star cluster and its notable planetary nebula.

while exposing your picture; but if you snooze, you may lose because you won't be around to interrupt your exposure if an airplane or satellite should streak into the field of view. Middle of the road guiding methods include using a video camera in place of an eyepiece or autoguider.

8.1 Using a Guide Scope

A separate guide scope permits you to use many of the same techniques you would use for piggyback photography. The principle is really the same, except that the physical size of the guide scope is usually smaller than that of the photographic telescope.

A separate guide scope can make it relatively easy to find a suitable guide in or near the area you are photographing. It also provides an easy way to guide directly on an object such as a comet which has a rapid proper motion. To set up a guide scope, you need only center the subject in your camera, look for a nearby guide star, independently point the guide scope at the star without moving the main telescope enough to decenter your subject, and securely lock the guide scope in place.

With all of these advantages, you may be wondering about disadvantages. The down side (life just wouldn't be interesting if there wasn't a down side!) is differential flexure between the guide scope and your main telescope. If too much flexure occurs, the result will be poorly tracked photos. The worst part is that if your telescope has a flexure problem, you probably won't realize it until you see your developed pictures! To minimize (and hopefully eliminate) flexure problems, a stable guide scope mount is important. Problems closely related to flexure include the shifting of internal optics in Cassegrain telescope which focuses by means of moving its primary mirror.

If you are guiding visually, the working focal length of your guide scope should be at least as long as that of the photographic optics. If instead you are using an autoguider, you may be able to use a shorter focal length guide scope. Some autoguiders do not work particularly well with slow f/ratio systems due to the relatively large Airy disk a slow f/ratio system produces. If your autoguider is picky about this, you may be able to use a telecompressor in the guide scope, but a better solution may be to get a larger aperture guide scope. A larger aperture guide scope will allow you to guide on fainter stars, making it easier to find a suitable guide star. It may also allow you to guide at a longer focal length, which will typically result in more accurate tracking.

If you are using a refractor or Newtonian reflector as a guide scope, it is important to use one with a relatively long original focal length. You can use a Barlow to increase the effective focal length of a guide scope to an acceptable degree, but reliance on too strong a Barlow lens can subject your system to flexure problems, particularly if the guide scope has a flimsy focuser like the type supplied on some small telescopes.

8.2 Off-Axis Guided Astrophotography

An off-axis guider, partial aperture guider, or beamsplitter permits you to guide on a star image which is formed by the same telescope that is used to take your pictures. This offers the advantage of virtually eliminating differential flexure, since the only flexure which could go undetected or result in false guiding corrections would have to be limited to the guider or your camera, and these items are not likely to pose flexure problems.

A beamsplitter will allow you to select a guide star that is actually within your photographic field of view. The disadvantage of a beamsplitter is that it attenuates part of the light from your telescope, increasing the required exposure time.

A partial aperture guider uses a reflector which is smaller than the local diameter of the light cone from the telescope. Most units will permit you to guide on a star within the picture area, and some even extend the range of guide star selection to include a zone immediately surrounding the picture area. The disadvantage is that the small reflector of a partial aperture guider will cast a penumbral (partial) shadow onto the film if it is used for guiding on a star within the photographic field of view.

An off-axis guider permits you to guide on a star which is in an area surrounding your subject, but not on one that is actually in the area you want to photograph. For deep sky photography, an off-axis guider is usually more desirable than a beamsplitter or partial aperture guider because it does not attenuate or obstruct any light from your subject. The full aperture of a telescope is then available for imaging your subject. The disadvantage is that the off-axis position of the guiding reflector limits the zone around a subject from which you can select guide stars. Fortunately, this difficulty is partially offset by the fact that an off-axis guider permits most or all of the telescope's available aperture to be utilized in imaging a guide star. This provides a brighter guide star image, permitting you to see dimmer guide stars.

When using an off-axis guider, finding and centering a guide star is not always as straightforward as it is when using a separate guide scope. With a guide scope, you just center the subject in your camera, pick a guide star, then point the guide scope at the star without moving the main telescope. Rare is the time you can set up this

Figure 8.2 (*opposite*). a An off-axis guider has a reflector which intercepts light from an off-axis guide star. **b** Rotating an off-axis guider on your telescope permits you to select a guide star from an area that encircles your subject. **c** This diagram shows where you would look for a guide star if the subject is centered and the guiding reflector (and usually the eyepiece holder) is directly above the field of view and you are previewing the picture area through an eyepiece positioned behind the guider. A typical radially adjustable off-axis guider permits you to select guide stars from an area corresponding to the narrow oval near the top of the drawing. By rotating the guider, a guide star can be selected from a "donut" having a thickness equal to the length of the thin oval. A typical guider having a fixed reflector only allows you to select a star from the center of the small circle at the top of the same thin oval. By rotating the guider, you can select from a thin ring of sky around your subject. The lower oval shows where to look for a guide star if you are looking into an eyepiece in the right-angle port of a combined multiple function flip mirror and off-axis guider such as the Versacorp DiaGuider. The dashed circle near the top of the drawing represents the field of a typical 24 mm eyepiece when the telescope is temporarily positioned so the subject is at one edge of the field of view. This position can be useful for judging the off–axis distance of a particular star. **d** Like most deep sky objects, the Dumbbell nebula (M27) has such a small angular size that its image through a typical amateur telescope fills only a small part of the 35 mm format. Such an object can be photographed even if you use a guide star inside the format area and the guiding reflector shadows part of the picture. This one hour exposure on ISO 400 slide film through a 20 cm f/10 Schmidt–Cassegrain was one of my first few deep sky astrophotos through a telescope, so I was not too picky about off center parts of the image. The absence of stars in the lower left reveals that I selected an easy to find guide star just inside the picture area. **e** A one hour exposure on hypered ISO 1600 color negative film through a 20 cm f/10 telescope provides an excellent image of M42, the great nebula in Orion. The large angular size and comparative brightness of this nebula make it relatively easy to photograph. This image has slight field rotation which is evident as increasingly elongated star images toward the top of the picture. The guide star was just outside the bottom of the picture. **f** A one hour exposure on off the shelf ISO 1600 film through the same telescope captures M17, the Omega nebula (also called the Check Mark nebula for obvious reasons). During this exposure, high wind at Stoneman Lake in Arizona vibrated my telescope so much that it caused slight streaking of the star images. The vibration even caused the guide star to look like a blurred streak in the eyepiece! **g** A large aperture, fast f/ratio Newtonian telescope can facilitate short exposure times. This image of the galaxy M33 required only a few minutes of exposure through a 37 cm f/3.8 Newtonian telescope. Pierre Schwaar built the telescope, equatorial mount, adjustable partial aperture "Piccadilly" guider (which permits guiding on axial or off-axis giude stars), and coma corrector that he used to take this picture. (Pierre's Piccadilly guider uses a small adjustable mirror which is affixed near the center of the diagonal mirror on Pierre's telescope. The mirror reflects light from a guide star to a second focuser on the side of the telescope.) Photo courtesy of Pierre Schwaar. **h** An exposure of only a few minutes through the same telescope imaged detail in the dust lanes of M31, the great galaxy in Andromeda. Photo courtesy of Pierre Schwaar.

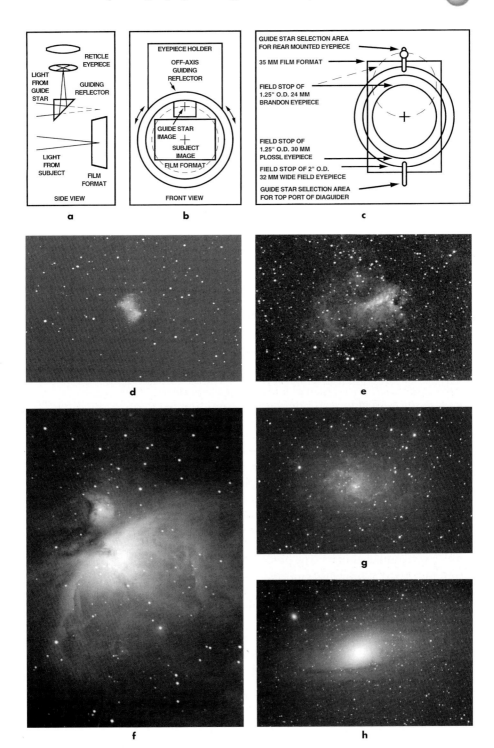

a SIDE VIEW

RETICLE EYEPIECE

LIGHT FROM GUIDE STAR

GUIDING REFLECTOR

LIGHT FROM SUBJECT

FILM FORMAT

b FRONT VIEW

EYEPIECE HOLDER

OFF-AXIS GUIDING REFLECTOR

GUIDE STAR IMAGE

SUBJECT IMAGE

FILM FORMAT

c

GUIDE STAR SELECTION AREA FOR REAR MOUNTED EYEPIECE

35 MM FILM FORMAT

FIELD STOP OF 1.25" O.D. 24 MM BRANDON EYEPIECE

FIELD STOP OF 1.25" O.D. 30 MM PLOSSL EYEPIECE

FIELD STOP OF 2" O.D. 32 MM WIDE FIELD EYEPIECE

GUIDE STAR SELECTION AREA FOR TOP PORT OF DIAGUIDER

d

e

f

g

h

quickly with an off-axis guider. Here, the guide star must to be close to a particular distance from the optical axis; usually somewhere between 16–20 mm. This means that if you want to center the subject in your photo, the guide star will have to be about this same distance from the subject at the focal surface. If you can't find a guide star in this area, you will have to decenter your subject in order to acquire an available star.

A radially adjustable off-axis guider can make it easier to find a guide star because it widens the area in which you can select a guide star to a relatively thick donut which surrounds your subject. Most radially adjustable guiders extend the guide star selection range to radii between of about 14 and 22 mm from the optical axis; however, the guider's eyepiece is not necessarily accessible from all of the possible rotary positions. In real world situations, only about half the donut's circumference can be comfortably viewed through most telescopes unless you use a relay diagonal.

Relatively few off-axis guiding systems have instructions thorough enough to enable first time users to get good results on their first try. The suggested step by step procedure for off-axis guided photography provided in this section will at least try to fill the gap. If you have a large aperture telescope or are using a separate guide scope (which is easier than off-axis guiding in many respects) you may still find the following information useful.

The following has been adapted from instruction manuals I wrote for using the patented and easy to use Versacorp VersaGuider off-axis guider and the DiaGuider and VersAgonal multiple function flip mirror attachments with Cassegrain and refracting telescopes. Most people do not have these particular accessories, so the following instructions have been modified to apply to more conventional guiders. The procedures are broken down into many steps. Once you are familiar with your equipment and the concepts behind the procedures, you may be able to skip or combine some of the steps.

8.2.1 Step by Step Guide to Prime Focus Off-Axis Guided Photography

1. Accurately polar align your telescope mount and start its sidereal drive.

2. Locate your subject in a low power (typically 20 mm focal length or longer) eyepiece and center it in the field.

3. Check the appropriate peripheral area around your subject (see Fig. 8.2c) for off-axis guide stars. If your eyepiece is mounted directly on your telescope, look above and to the sides of your subject for a guide star. If you are using a star diagonal, look for guide stars below and to the sides. Note the rotary position of the diagonal attachment (if used) and the location of prospective guide star(s) with respect to your subject. Once you have located at least one prospective guide star, note the appearance of your subject, star patterns, etc., in both your finder scope and camera viewfinder so you can find your way back to the subject after the next few steps.

Remarks: The area searched for a guide star should be considerably outside an area which corresponds to the short dimension of your camera's image format. If you are using a 35 mm camera, you should look for a guide star that is considerably more than 12 mm from the center of the image. This is well outside the field of even a low power 24.5 mm (0.965 inch) OD eyepiece and at or slightly outside the edge of a typical low power 31.7 mm (1.25 inch) OD eyepiece field. Shadowing from the guiding reflector is minimized when it is positioned so its shadow is just outside one of the long sides of your format.

If the eyepiece you use to find your subject is mounted directly on your telescope or immediately behind your guider, the "top" of the field in subsequent steps will be the true top of the eyepiece field.

If instead you are using the eyepiece in the top of a star diagonal or right-angle flip mirror attachment, refer to the part of the field toward your telescope as the "top" in subsequent steps. In addition, the orientation of the eyepiece holder on your star diagonal will be considered the 12 o'clock position for the eyepiece stalk on your guider. If you rotate your diagonal until the prospective guide star appears to be toward the bottom of the field with respect to your subject (with the bottom being the part of the field away from your telescope tube), you can usually acquire the star in your guider by simply orienting its eyepiece stalk (and thereby its off-axis reflector) in the same rotary position as that of your star diagonal.

These directions assume that your guiding eyepiece will be mounted perpendicular to your camera, and that you can most easily access it from above or to either side of the guider. If your guiding eyepiece is easier to access from a different angle, then emphasize looking for a guide star on the side of your subject toward the guiding reflector, while considering positions that provide the most comfortable eyepiece position, as seen from the back of the guider. If you will be using a relay diagonal with your guider, you also want to emphasize looking for a guide star on the side of your subject that is toward the guiding reflector.

The object is to intercept a guide star with your off-axis guiding reflector while the image of your subject is centered in your camera.

4. Attach a guider with a reticle eyepiece installed to your telescope, then attach your camera and counterweight. Verify that the reticle eyepiece is positioned to provide a properly focused star image.

Remarks: If the reticle eyepiece is not positioned for proper focus, loosen its lock screw, move it in or out of your guider to set the correct position, then tighten its lock screw. This position is established after you perform step 8 for the first time. If you are using a particular guider or eyepiece for the first time, you can perform step 8 on a bright star, the moon, or a terrestrial object prior to your deep sky photo session.

5. Point your telescope at a sufficiently bright star within a few degrees of your subject and focus its image in your camera. (Or you can use a focusing attachment.) *Do not* adjust your telescope's focus control after you have focused the star image in your camera.

Remarks: If your camera focusing screen does not have a matte or clear center aerial central area, focus the star on the plain matte part of the screen, just to the right or left of any central split image spot, etc., it may have. If you can find a double star having sufficient separation to see in your camera viewfinder, focus on it. This can be easier than focusing on a single star.

6. Move the telescope back to your subject and center the subject in the viewfinder.

Remarks: If your subject is too dim to see in your camera finder, it is desirable to parfocalize your eyepiece and camera. This can be done with a suitable extension tube or some star diagonals if you move your eyepiece in or out just enough that the eyepiece and extension tube or diagonal assembly can be interchanged with your camera without refocusing. If you don't have the right gadgets to parfocalize your eyepiece and camera, you may instead be able to mount or simply hold a low power eyepiece the appropriate distance behind your guider body (with the camera removed) to see your photographic field. Once your subject is centered, you can reattach your camera.

Multiple function accessory systems such as the patented Versacorp DiaGuider (a little plug there!) offer the convenience of simultaneous attachment of a camera and low power eyepiece. With this combined flip mirror and guider, you can perform parts of steps 4 to 6 before steps 2 and 3. This will keep you from having to reacquire your subject in your camera finder and make locating and centering your subject a lot easier. If you are also using the Versacorp MicroStar™ focusing attachment in the top port of the DiaGuider™, you can use it to focus without removing your camera; or even before you attach your camera for that matter.

7. Using one of the procedures below, acquire the guide star you recall from step 3 in the reticle eyepiece on your guider. (Or you can instead acquire it with a parfocalized low power eyepiece which is temporarily used on your guider.) The object is to rotate the guider on your telescope to a position which will allow its guiding reflector to intercept the guide star you noted in step 3 yet not move the telescope enough to excessively decenter your subject.

Remarks: The rotary guider position you should use will depend on how your low power eyepiece was attached in step 3. If the eyepiece was mounted directly on your telescope or immediately behind your guider, simply rotate the guider to a position that will cause its reflector (and usually its eyepiece holder) to be in the same rotary position your guide star appeared to be from the subject in step 3. (In other words, simply rotate the guider so that its top points in the same direction as the guide star was from your subject in your low power eyepiece.) For example, if the guide star was at a one o'clock position with respect to your subject, rotate the guider so its reflector is also at the one o'clock position. The eyepiece stalk of an off-axis guider will usually have the same rotary position as the guiding reflector.

If instead you had used the eyepiece in the top of a star diagonal or right-angle flip mirror, the orientation of the top of the guider as seen from the rear should be a "mirror image" (with the "mirror" horizontal) of the orientation of the star to your subject; i.e. if the guide star is at the 5 o'clock position below your subject, the top of the guider should point toward the 1 o'clock position as seen from the rear. For review if you use a star diagonal when you locate your subject, the part of the eyepiece field toward the front of the telescope is always considered to be the top, and the previous orientation of your diagonal is always considered to be the 12 o'clock position for the guider.

As mentioned in step 3, if your diagonal was rotated until the prospective guide star appeared to be toward the bottom of the field (i.e. the 6 o'clock position, which was the part of the field away from your telescope) with respect to your subject, you can usually acquire the star in your guider by simply orienting its eyepiece stalk in the same rotary position as that of your star diagonal (which would correspond to a 12 o'clock position).

If the subject and guide star are bright enough to see in your camera finder, you can orient your camera and guider so that you can use them to make it easier to acquire the guide star in your guiding eyepiece. With your subject centered, rotate your camera so the long dimension of your picture is vertical with respect to the guider, then look for a guide star near the bottom edge of your finder (the side toward the bottom of the guider). Keeping the subject centered, rotate the guider with your camera until the star is positioned at the middle of the bottom edge of your camera finder. If your camera has a right-angle finder, center the star on the top edge. The shadow of the guiding prism may prevent you from positioning the star at the very edge of your camera finder. If so, slightly move your telescope to move the star image toward the edge of the screen until it dims considerably.

If a guide star is not visible in your reticle eyepiece after you perform the above steps, slowly rotate your guider up to a few dozen degrees until a guide star is found and roughly centered from side to side. You can move your telescope slightly to acquire the guide star and position it vertically but too much motion (enough to move the star more than about $\frac{1}{4}$ of your reticle eyepiece field) will noticeable decenter your subject.

If your guider has radial adjustment (as do Lumicon and Versacorp guiding systems) use this feature instead of moving your telescope. This will allow you to keep your subject centered. If the star still is not visible after this, verify that your telescope is pointed at your subject, then look for the star again.

8. Once the star is in your reticle eyepiece field, loosen the guider's eyepiece holder lock screw and move the reticle eyepiece up or down until the star appears to be in relatively good focus, then lock the eyepiece in place.

Remarks: To save time in the future, you can mark the barrel of your eyepiece (or use a stop ring) so you can roughly position it for proper focus without using a star.

If you are using an autoguider, attach it at this point, focus it, then go to step 12. When you focus the star image in an autoguider at this point, do so by sliding the autoguider in and out of the guider's eyepiece holder. You do not want to adjust the telescope's focuser at this point because doing so would change the focus at your camera's focal plane.

9. Turn on your illuminated reticle eyepiece and verify that the reticle lines look sharp at least in the center. If they do not look sharp, focus the top part of the eyepiece (if your eyepiece can be focused in this way) until they do. Once the lines look sharp, set the illuminated reticle brightness so the reticle lines are as dim as you can get them without making them hard to see.

10. Center the guide star in your reticle eyepiece, then verify that the reticle lines in your eyepiece are roughly parallel with the motion of the star when the telescope is moved in RA and declination. If the lines are not at least within a few degrees of being parallel, loosen the guider's eyepiece holder lock screw and rotate the eyepiece until the lines are roughly parallel to the star motion, then lock the eyepiece in place.

Remarks: If your guider is rotated sideways to acquire a guide star, the telescope motions used to move the star image will typically be reversed from the usual orientation; i.e. RA will now move the image up and down (if down is considered to be toward the back of the guider) and declination will now move the image from side to side.

If more than one star is available for guiding, use the one that will allow you to assume the most comfortable guiding position. If you want to use a visual Barlow lens in your guider, attach it at this point. A Barlow lens will usually permit you to guide more accurately.

11. With the guide star centered, move your head slightly from side to side while still looking at the star image in the eyepiece. If the star seems to

move from side to side in relation to the reticle, adjust the focus by slightly moving your reticle eyepiece up or down until no motion is seen.

Remarks: This step is to ensure more precise focusing of the guide star image. The plane of best focus should be coincident with your reticle so there will be no any parallax which could cause you to make false guiding corrections!

12. Rotate the camera to the desired angle for picture composition, accounting for any shadowing by the guiding reflector.
13. Check to see that all lock screws, etc., on your telescope and guider are tight.
14. Attach a locking cable release to your camera, set the speed to "B", and wind the shutter. If your camera has a "T" setting, a cable release may not be required.
15. Be sure you are in a comfortable guiding position, taking into account where the eyepiece will have moved to by the end of your exposure.

Remarks: It is also important to check that you are not in a position where you could accidentally bump or touch your camera or reticle eyepiece during your exposure. Just the slightest tap can cause some bayonet mount cameras to move enough to cause visible streaking in the image. It may not be a bad idea to deliberately but lightly "pre-bump" your camera before you start your exposure.

16. Precisely center your guide star in the eyepiece reticle.
17. Stand up, stretch out, relax, and check the sky for picture killing airplanes. If you are out alone, get your CD or tape player ready. If you are in a group, show some friends the part of the sky you will be photographing and see if they will check the area for airplanes and bright satellites every now and then.
18. Precisely center the guide star again if necessary. Take a break to rest your eyes.

Remarks: It is important to determine whether the star's drift is excessive. If it drifted a substantial distance in right ascension, your drive may not be turned on, or it may be set for the wrong hemisphere. If the star visibly drifted in declination, your telescope may not be adequately polar aligned. By noting any significant drift in declination, you can tell at least something about how well your telescope mount is polar aligned. For more information, see data on the "drift method" of polar alignment in Chapter 7.

19. While observing the centred guide star image in your eyepiece, open the camera shutter. (Then open your guider's manual shutter if it has one.)
20. Guide your photograph carefully! Look up occasionally to rest your eyes and look for airplanes or satellites which may move in front of the area you are photographing.

Remarks: If you are using a conventional 12 mm double line illuminated reticle eyepiece in which the reticle lines are $\frac{1}{5}$ mm apart, you will need to guide to an accuracy of considerably less than $\frac{1}{4}$ ($\frac{1}{2}$ if you are using a guiding Barlow) of the width of the reticle's central box to get a really good picture. (The required guiding accuracy can be at least twice this critical if you are using a CCD imager.) If there are bright stars in your picture, a guiding error of only one second duration could cause these stars to appear streaked or elongated.

Some guiding systems (such as those offered by Versacorp and Questar) permit you to interrupt your exposure while still permitting you to see your guide star. This feature will permit you to take a break during your exposure if necessary, or to break a long exposure up into a number of shorter guiding sessions.

21. Close camera shutter.

Remarks: It is important to stick with your predetermined exposure time. If you are not in the mood for guiding, your brain may play tricks on you. As I have guided photos, thoughts have come into my mind like: "It looked so bright in the eyepiece, the picture will surely be *acceptable* (so much for a *good* picture!) with a shorter exposure – and I'm cold, so the emulsion is cooled, right?"

Wrong! Don't give in and shorten your exposure. Go the distance, and you will be glad you did! Remember, an underexposed photo is a spoiled photo!

If you make guiding errors well into your first exposures or discover that your polar alignment is slightly off because of frequent declination adjustments in the same direction, continue to guide the rest of your exposure anyway. Even if an error causes the star images to be streaked, you will still be rewarded with a good image of your subject and begin to discover the acceptable amount of guiding error for your applications.

Good luck, and may all your star images be round!

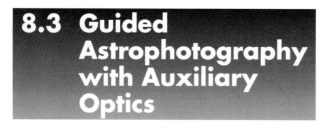

8.3 Guided Astrophotography with Auxiliary Optics

Auxiliary optics can be used to change the effective focal length of your telescope and permit you to get different image sizes. Chapters 2–4 cover some specific types of auxiliary optics. This section will provide a brief review and some additional observations.

A telecompressor (also called a focal reducer) provides a smaller image size and a Barlow lens or tele-converter provides a larger image size. Eyepiece projection can provide a still larger image size, but the resulting f/ratio is so slow that this technique is most often reserved for bright subjects such as the sun, moon, and planets.

Figure 8.3. Good images of showpiece objects such as the Omega Centauri globular star cluster can be obtained through a suitable moderate size telescope even when the image scale is reduced with a telecompressor. This photo was taken on unhypered ISO 1000 slide film through a 20 cm f/10 Schmidt–Cassegrain telescope. A 0.7× telecompressor shortened the effective focal length to 1400 mm and provided an f/ratio of f/7, shortening the required exposure to a mere 20 minutes.

The degree of effect a given auxiliary optic will have is almost always determined by its distance from the focal surface; a greater distance results in more reduction for a telecompressor or more magnification for a Barlow lens. Some auxiliary optics are optimized for use at a prescribed distance from the focal surface, but most can be used throughout a reasonable range of distances from the focal surface with minimal effect on the image quality.

If you use an auxiliary optic behind a guider or flip mirror, the eyepiece in the top port of the guider or flip mirror will have to be moved up or down if it is to be parfocal with your camera. (Parfocal means that the same focus setting works for both the eyepiece and camera.) If you add a telecompressor behind a guider, then the eyepiece in the top port will have to be moved up a few centimeters. (Extension tubes are usually available for this.) If you add a Barlow lens, the top eyepiece (if it has to be moved at all) will typically have to be moved down. If you add an eyepiece in back for projection photography, the top eyepiece position will be dependent on the focal length and longitudinal position of the projection eyepiece. If instead an auxiliary optic is used in front of the flip mirror, the relative top eyepiece position probably won't have to be changed.

8.3.1 A Telecompressor for a Smaller Image

A telecompressor is most often used to provide a faster f/ratio for a given telescope. This can be desirable because a faster f/ratio will facilitate a shorter exposure time; however, a tradeoff is that the image scale will be reduced. Reduced image scale does not mean that you will get a substantially wider field of view from a given telescope when you use a telecompressor; rather, the telecompressor simply reduces the circle of coverage a telescope already has down to a smaller size.

This is not to say that a telecompressor can't be used to provide a wider field *on a particular film format*. In cases where the telescope covers a lot more area than a given film format, a suitable telecompressor can provide a wider field of coverage *within that format*; however, this will not usually be a wider absolute field of view than what you would get by simply using a larger format camera without a telecompressor.

A telecompressor reduces the illuminated field of a telescope by about the same degree as the reduction in image scale it provides. This relationship is by no means exact because the resulting illuminated circle size can vary according to the aperture and longitudinal position of a telecompressor. A telecompressor having a small lens typically provides a much smaller image circle than its reduction value would indicate, resulting in an actual reduction in a telescope's photographic field of view. Some small telecompressor lenses will completely vignette all but about the central half of the 35 mm format.

Less vignetting will result if a larger telecompressor lens is used, but there is a limit to how large a lens diameter is practical or even necessary. A large lens of a given design and optical power will tend to have more aberrations than a small one. It will also cost more and weigh more. In addition, there is little point in using a lens which is larger than the original illuminated circle of the telescope. If the lens is larger than necessary, its outer zones simply won't be utilized.

For instance, a typical 20 cm f/10 Schmidt–Cassegrain telescope has a rear aperture of 38 mm, so there is little point in using a telecompressor with much more entrance aperture than this if it will be positioned right up against the rear cell of the telescope. If the telecompressor is instead mounted in a sufficiently large cell and positioned several centimeters behind the telescope, it may be possible to get some additional benefit from a lens up to 50 mm in diameter. An 80 mm diameter telecompressor lens would be overkill for such a telescope, though it may be fine for a telescope having a relatively large exit aperture.

The distance between a given telecompressor and the focal surface will determine its coverage with a particular telescope and how much back focus will be required. Distance from the focal surface will also influence the magnification of a given telecompressor lens and how feathered the edge of its coverage will be. For a given magnification, a relatively strong telecompressor lens positioned near the focal surface will provide more even illumination in the central $\frac{2}{3}$ or more of the image circle than would a relatively weak lens used at a longer distance from the focal surface. The edge of coverage for the strong lens will be rather sharply defined. The weak lens provides full illumination over a smaller area, but the off-axis image will darken more gradually toward the edge of the covered field.

It used to be that the optics in telecompressors were simple doublets which telescope makers already used in other products. One example is the small 30 mm aperture 0.5× T-thread telecompressor that was popular many years back. Its optics are the same design as the objective lens in a short 6 × 30 finder scope offered by the same manufacturers. These older telecompressor lenses work to a degree, but they offer no correction of field curvature or off-axis aberrations. In fact, many introduce small aberrations of their own. In the last decade or so, a few manufacturers have introduced telecompressors and "reducer/correctors" which are optimized specifically for their intended use. Some of these are even matched to particular telescopes.

Exposure compensation for a telecompressor has a lot to do with its off-axis illumination characteristics and how much of the field is occupied by your subject. If your subject fills only the central third of the field, then a telecompressor will reduce the required exposure by about the square of its reduction value. Therefore, a 0.5× telecompressor used in this context will allow you to shorten the exposure time by a factor of four. If instead the image from your telecompressor has an obvious reduction in brightness at a moderate off-axis distance and your subject fills most of the illuminated field, you may have to use an exposure time that is at least 20 percent longer than what you would expect from the reduction value. In this case, the center of the image may be noticeably brighter than an area more than halfway to the edge of the field.

8.3.2 A Barlow Lens for a Larger Image

If you want to get a large image of a deep sky or other object, a Barlow lens or photographic teleconverter will usually do the trick. This will result in a slower f/ratio from a given telescope, but you can compensate by using a longer exposure.

Exposure compensation for a Barlow lens is equal to the square of its magnification, so if a Barlow works at a magnification of 2×, it will double the focal length of your telescope and you will need to increase your exposure time by a factor of four (2 × 2). If your film has severe reciprocity failure, it may be necessary to increase the exposure even more.

In order to minimize the potential for flexure problems during your exposure, it is typically best to use a Barlow lens in front of your guider, provided you can do so and still get the desired magnification.

Unlike telecompressors, Barlow lenses rarely cause reduced illumination off-axis. If they do, it is usually because the lens is too small or a baffle is obstructing part of the off-axis light. You can tell if a Barlow lens is large enough by attaching your empty camera to the optical system, opening the back and shutter, then looking through a corner of the shutter opening from a distance. If the full aperture of your telescope is visible through the corner of the shutter opening, then the Barlow lens is large enough. If some or all of the aperture is obscured, the next step is to evaluate whether the obscuration is caused by the edge of the Barlow lens. Most often, the obstruction will be a baffle ring of some kind. Some obstruction is allowable, but it is best if you can see at least half of the telescope's aperture from the corner of the 35 mm format. A 2× or stronger Barlow lens of 2.5 cm aperture will usually illuminate the entire 35 mm format.

Simple Barlow lenses can work quite well for photography with relatively slow f/ratio telescopes, but faster telescopes usually require more sophisticated optics. Here, a 2× photographic teleconverter (particularly a well corrected one like the Nikon TC-301) can radically outperform a typical astronomical Barlow lens. One of the primary problems with simple Barlow lenses is curvature of the focal surface, which becomes more of a problem with fast f/ratio optics due to the steeper convergence of the light cone. In general, a weak Barlow lens used at a relatively long distance from the focal surface will cause less field curvature than a strong lens of the same complexity which provides the same magnification when used at a close distance.

8.4 Guided Cometary Photography

Comets are interesting and sometimes beautiful objects which people photograph for both aesthetic and scientific purposes. In research, photographs have been used both to discover comets and to refine their orbital data. Since visible comets are inside our solar system, they can get close enough to the sun to have a relatively fast orbital velocity. Therefore, it is not unusual for a comet to have a relatively rapid motion (called proper motion) in relation to background stars. In such a case, appropriate guiding techniques are required to compensate for this motion and get good detail in images of the comet itself.

Shortly after acquisition or discovery, it is not unusual for a comet to be dimmer than most charted stars

Figure 8.4. Top left: A short 10 minute exposure of comet Halley (boxed) through an 85 mm lens reveals no apparent streaking, but the image scale is too small to show any detail in the subject. The comet was close to magnitude 7 at the time, just barely detectable with the unaided eye. This cropped image is enlarged about 3×. The small rectangle at center shows the field of view for the right image. Top right: This one hour exposure of comet Halley (which was dim at the time) was taken through a 20 cm f/10 Schmidt–Cassegrain telescope and guided on a background star. The visual appearance of the comet was more or less round at the time, but its motion during the exposure caused its image to be badly streaked. This is no surprise because the larger image provided by a telescope enlarges both a subject's detail and the effects of its relative angular motion. If the image had instead been guided on the comet, it would be relatively sharp and all of the background stars would be streaked. Bottom left: A background star was used as the guiding reference for this 5 minute exposure of comet Hyakutake on ISO 800 color negative film through a 300 mm f/4.5 ED Nikkor lens. Bottom right: The same equipment was used for this 10 minute exposure which was guided on the comet's nucleus. This provided a far sharper image of the comet's coma and tail, but it also caused the background stars to appear streaked.

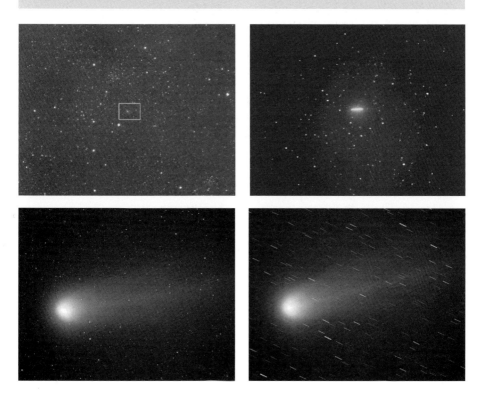

in the surrounding part of the sky. A long exposure is required to image such a dim comet, but the comet's motion may (depending on the rate of motion) cause its image to be streaked if one guides on a star in the usual way. A comet (particularly a dim one) is a great deal easier to see in a photo if it is imaged sharply and the background stars are allowed to appear streaked. If a comet is bright enough, one can often track it by guiding on the nucleus with the use of a beamsplitter, partial aperture guider, or separate guide scope.

If the comet is too dim to see and guide on with a guide scope, or if you are using an off-axis guider, you may need to use an offset guiding technique. This is where you guide on a star, but you also compensate for the comet's motion. Proper offset guiding requires advance knowledge about the comet's rate and direction of motion. Once this is known, it is possible to guide on a star and use a translating eyepiece or reticle in which the correct lateral motion is imparted by a calibrated knob or motorized control. Effective offset correction can also be provided by translating a relay lens or Barlow lens.

A simpler offset guiding method from an equipment standpoint is to use a reticle having several reference lines along one axis and guiding in a way that causes the star image to gradually drift from one reference line to the next over a predetermined period of time. Offset guiding can also be used for asteroids and other objects having a known rate and direction of proper motion.

Cometary photography can be rewarding and result in beautiful images, and while your photograph probably won't be the one in which a comet is discovered, you never know if it will turn out to be useful for research. Some comets are so dim that they can only be imaged with a large telescope, while others are so bright they can be photographed with a mere camera on a tripod. The main thing is that you can photograph at least some comets – and have fun doing it!

Chapter 9
Photographing Astronomical Events

Astronomical events are a special category of subject. This is when the sun, moon, planets, and certain stars put on a dynamic show right before your very eyes! These dramatic and interesting events include meteor showers, novae and supernovae, conjunctions of the moon and planets, occultations of stars or other objects by asteroids, major planets, or the moon; and eclipses of the sun and moon.

Some events may last several days, while others may last only seconds. Either way, an astronomical event occurs at a given time and offers no second chance for the astrophotographer. Once an astronomical event is over, it's really over, never to be repeated again in exactly the same way.

If you want to get good pictures of a short lived event, you have to get it right the first time. This requirement for excellence applies not only to astronomical events, but also to photographing other transient phenomena such as comets, variable stars, or earth related events such as aurora displays, satellites, rocket launches, air shows, and illusive types of wildlife.

Before you photograph an astronomical event, it is a good idea to establish some goals and determine which ones you really want to emphasize. Other aspects of your photography should be considered discretionary so you don't try to do too much. This organized approach is useful for any type of observing or photography, but it can be essential when photographing astronomical events.

If you want to photograph impressive astronomical events and have some hope of being calm at the same

Figure 9.1. Top left: A camera and normal 50 mm lens on a fixed tripod was used to photograph a rare aurora display (visible below cloud) that was visible from as far south as Colorado. Top right: A wide angle lens and fast film can capture bright events during a meteor shower. Richard Payne of Avondale, Arizona captured several bright 1998 Leonid meteors streaking past a rare glowing meteor trail remnant in this 15 minute exposure on ISO 3200 black and white negative film through a 16 mm f/2.8 fisheye Nikkor lens. This 6× enlargement includes only a small portion of the format. Photo courtesy of Richard Payne. Bottom left: An inexpensive 400 mm lens on a fixed tripod was used to photograph the partially eclipsed moon as it set behind the Rocky Mountains. A lunar eclipse proceeds at a leisurely pace and partial phases in particular can be photographed with modest equipment. This cropped 3.5× enlargement shows the field of view you would get with a 700 mm lens. Bottom right: Photographing events which occur on other planets can require more sophisticated equipment than the other subjects. David Charles of Loveland Colorado used eyepiece projection with a 20 cm f/10 Schmidt–Cassegrain telescope to capture the aftermath of the 1994 collision of Comet Shoemaker–Levy 9 with Jupiter. Photo courtesy of David L. Charles.

time, it is helpful to become so familiar with your equipment and procedures that you can take your pictures while barely having to think about it. Such familiarity can usually be accomplished with sufficient practice. More particularly, it is good to practice with the exact equipment setup you plan to use when you photograph a given event.

Generally, procedures such as packing, unpacking, and setting up equipment need only be practiced a few times; but the actual procedures which must be followed during and within a few minutes of the astronomical event itself may have to be practiced ten times or more if you are to ensure that you and your equipment will perform adequately. This is particularly true if you try to operate several cameras during a short lived event such as the total phase of a total solar eclipse. Circumstances at the time of the event may cause you to be tired while in unfamiliar surroundings, but familiarity with your procedures can help lessen the effect this may have on your results.

Each time you practice, it is a good idea to make a list of things you can do to improve your setup, then make as many improvements as you can before your next practice session. Ultimately, you want to develop and document a specific procedure for any event which lasts only a few minutes or seconds. Ideally, such documentation for a really special short duration event should list the exact second you want to take almost every exposure and make every observation. It is important to allow adequate time for observation, because you won't know how your pictures should look if you fail to observe the event you photograph!

Other useful documentation can include a list of things to do before you go to your observing site and a checklist which covers everything you want to bring with you. It is best if your list of what to bring includes even seemingly obvious items like extra clothing and the like. If getting to your site at a reasonable time means that you have to rush out of town right after you get off work, you don't want to leave any part of your trip to mere memory. It is good to have a specific plan. You can always improvise if necessary. In addition to your list, you can make an audio or video tape to play during the event and remind you to perform various steps.

All of this planning is obviously dependent on the characteristics of each event you observe and photograph. Good sources of information about astronomical events include periodicals such as *Astronomy* or *Sky & Telescope* magazine. Other sources include books having tables of upcoming eclipses and the like. The Internet can also be a good source of information once you locate the sites which provide reliable data.

Astronomical software can be the best source of all when it comes to getting general data about events such as conjunctions of major solar system objects because you can use it to find out about an upcoming event well before the time it is covered in a magazine.

In addition, appropriate astronomical software can show what events such as occultations by the moon will look like from your particular location.

9.1 Meteor Showers

Meteor showers are unique photographic subjects because photographs seldom capture them in the way they are observed. This is partly because a meteor moves rapidly across the sky and is visible for only a fraction of a second to several seconds. The brightness can range from so faint as to be invisible, to a rare fireball, or bolide, which can be brighter than the full moon. Visual observers can watch a meteor streak across the sky, and in some cases, even see a faint trail which a meteor may leave behind. Video may show a bright meteor in motion, but it will not record dim ones very well.

A film image of a meteor usually requires an exposure time at least as long as the brief period that a meteor is visible. Since the exposure is so long, the meteor is recorded as a streak. Even though a film image will not capture real time motion, it offers the advantage of recording the actual path of the meteor, which is useful for both aesthetic and scientific purposes. An additional advantage to film photography is that a long exposure time can allow more than one meteor to be captured on a single frame.

Meteors are not actually called meteors until their final seconds, when they reach the atmosphere and burn up, going out in a blaze of glory, as it were. If a meteor should hit the ground, it would then be called a meteorite. Before they reach the earth's atmosphere, meteors are just small, unobservable spaceborne particles of debris (meteoroids)which drift around the solar system. Some debris clouds which produce meteors are widely scattered comet remnants which orbit the sun in more or less the same orbit that the comet had occupied. These tend to provide meteor showers at predictable times every year when the earth's orbit causes it to encounter the related debris cloud. Some regular meteor showers are named after comets which occupy the same orbit around the sun as the debris cloud. Other debris clouds have different origins than comets.

Most annual meteor showers are named after constellations from which the related meteors appear to radiate. The meteors obviously don't originate from the constellation; rather, they appear to radiate from it due to the motion and orientation of the earth relative to particles in the debris cloud. It is not unusual for the radiant point to rise at about midnight, since this is the time an observer is at the transition between the leading and trailing parts of the earth. At this time, it is not unusual for meteors to move across the sky in a predominantly east to west direction.

Meteor showers are almost always best after local midnight because the observer is then on the leading side of the earth as it moves in its orbit. The leading side of the earth encounters more meteors than the leeward side. This can be illustrated in familiar terms when you consider the fact that you are likely get more water on your face than on your back when you run through the rain.

To photograph meteors, all you need is a camera capable of time exposures and some stable means of pointing it at the part of the sky you want to photograph. It is preferable if you have a wide angle lens of moderately fast f/ratio (f/2.8 or faster) since this will increase your chances of capturing meteor images. It is not necessary to track the stars when you photograph meteors, though a tracked photo will usually be more reminiscent of the event as you saw it. Stars appear streaked in an untracked photo, while they look more normal in a tracked one.

It is not unusual for most photos of an average meteor shower to show no meteors at all. Meteors are often dimmer than they seem, and their rapid motion across the sky results in a short effective exposure time. Even if a meteor is visible for several seconds, it is only in a given point in the sky (and at a given point on your film) for an instant. A fast film and f/ratio can help record meteors, but this will limit the exposure time you can use before your picture reaches the sky fog limit, brightening the sky background in your picture. If a meteor event occurs during your short exposure with a fast film and fast f/ratio, its image will be brighter than it would have been with a slower film, though it will usually be more grainy. If you will be making a composite of several pictures, the best bet may be to shoot the background sky with slow film, then shoot a lot of short meteor exposures with fast film. After this, you can combine the images through composite printing or other means. This will give you both a sharp sky background and bright meteor images.

Using a shorter focal length lens to widen the angle of view will increase how much sky you can cover in one picture, but it will also reduce light grasp – even if the f/ratio remains the same. As mentioned in chapter 4, the true aperture of your optics can be determined by looking into the front from a distance. We can see what can happen to your aperture when you reduce focal length by examining the extreme case of an all-sky reflector. The effect of the reflector works both ways; it reduces the image scale to widen the field of view, but it also reduces the apparent size of the aperture when you look at the reflection of your camera in the reflector.

A 50 mm f/2 lens such as you may use to photograph an all-sky reflector has an aperture of 25 mm, but the all-sky reflector reduces the effective focal length to about 7.5 mm. Since the f/ratio is still f/2, the aperture is only 3.75 mm. This is about 50 times less aperture area than that of your 50 mm lens! Accordingly, your 50 mm lens alone will photograph a meteor up to about 50 times (over 4 magnitudes) dimmer than it will when used with the all-sky reflector. Similar effects on limiting magnitude are true of wide angle lenses.

Aperture area is what determines light grasp, excluding factors such as transmission value. Aperture is dependent on focal length and f/ratio regardless of what type of optics you use. A 16 mm f/2.8 fisheye lens has an aperture of about 6 mm, which is an improvement over the previous all sky refractor example. Most 16 mm fisheye lenses are very sharp, and one of them can cover the full sky if its lens hood is removed and it is used with a 6 × 6 cm film back.

The above dilemma may make photographing dimmer meteors seem like a lost cause, but there are ways to image them. One way is to use an image intensifier. Another is to use a bank of cameras with relatively large aperture lenses of a moderate angle of view to simultaneously photograph a mosaic of adjoining areas in the sky.

Whatever equipment you use, the best maximum exposure for meteor photography from a dark site is usually about the same as what you would use for ordinary "deep sky" piggyback photos.

9.2 Conjunctions

Conjunctions can be observed and photographed with the simplest of equipment. The best views are often

Figure 9.2. Top: The 15 June, 1991 quadruple conjunction of the moon (left), Venus (right), Mars (top) and Jupiter (center) was photographed on Kodachrome 64 film with a Tokina 50–250 mm f/4–5.6 zoom lens which was set to a focal lenght of about 180mm. The exact exposure time was not recorded, but it was probably about 20 seconds at f/5. The enlargement is 4.5×. Below left: A long exposure reveals earthshine on the moon during its 26 October 1984 conjunction with Venus. A 20 cm f/10 SCT and a 0.7× telecompressor were used for this 30 second exposure on ISO 1600 film. Below right: On rare occasions, the angular distance between planets can be so small that they both fit in the area covered by a telescope. This 23 February 1999 conjunction of Venus and Jupiter was photographed on ISO 1000 color negative film with a 9.4 cm f/7 refractor and a low magnification Barlow lens which was stacked with a 2× teleconverter. The exposure was about 1.5 seconds at f/20. The image is enlarged about 4.5×.

with your unaided eyes, so really good photos can be taken with a camera having a normal or moderate telephoto lens. If you take a conjunction picture at twilight, you may even be able to get a good shot with an automatic camera. Close up pictures of tighter conjunctions can be photographed with a telephoto lens or even a telescope. In the latter case, your picture may even show details such as the moons of Jupiter or even the phase of an inner planet.

Some conjunctions consist of just one planet or star in the same general part of the sky as the moon. Others may involve many objects. Those which involve two or more planets may even occur in a series, particularly when the outer planets move in a retrograde fashion. Retrograde motion of planets outside the earth's orbit occurs when the planet is near opposition, or its closest approach to the earth. This is also the time when the earth passes more or less between the planet and the sun. (I say more or less because an exact lineup in all axes would be very rare.) All planets outside the earth's orbit have a slower orbital motion than the earth, so when the earth passes by, its faster motion makes the slower planet appear to move backwards (east to west instead of the normal west to east) in relation to the stars. This is easier to see when you consider that if you view the solar system from the north, the planets will appear to move around the sun in a counterclockwise direction. The earth also appears to rotate counter-clockwise when viewed from the north.

9.3 Occultations

Occultations of planets or stars are among the most dramatic of astronomical events short of a lunar or solar eclipse. Here, one celestial object actually covers another. Occultations are not as common as conjunctions, since an occultation requires an exact lineup, or "syzygy", of the observer and the celestial objects involved.

9.3.1 Occultations by the Moon

Shortly before an occultation of a star or planet by the moon, the subject appears to be in conjunction, except that the object to be occulted is east of the moon rather than being to the north or south of it. As the moon draws closer to the object, it becomes more obvious that something special is going to happen; the moon is going to cover the object!

When the occulted object is a major planet, it will appear to go out gradually as it is covered. If instead it is a star, it will appear to go out instantly. If it is a double star, it will typically appear to go out in stages. When an object emerges from behind the moon, it typically appears to do so in the same general way as it disappeared, though the emergence of a double star may vary according to the orientation of its components to the local limb of the moon.

If the moon is a young crescent, the occultation can be more spectacular because the occulted object will first be covered by the darkened part of the moon. This

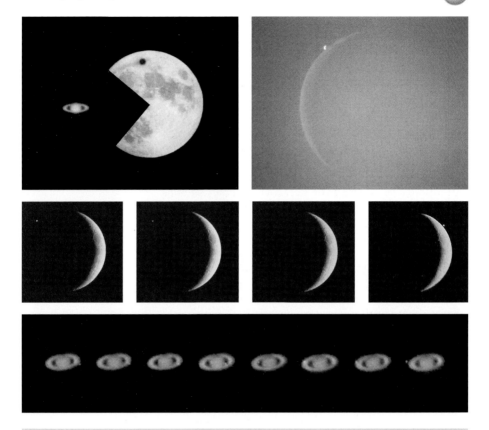

Figure 9.3. Upper Left: A hapless planet is occulted by a fictional "Pac Moon". Upper Right: Events such as this 9 July, 1980 occultation of Venus by the old crescent moon can be photographed in broad daylight. This photo was taken with a 9 cm f/11 Maksutov–Cassegrain telescope and a Vivitar 2× teleconverter. Middle row: The same telescope and teleconverter were used to photograph this 15 July, 1980 occultation of Jupiter by the moon. The left 3 images which show Jupiter being covered are exposed for $\frac{1}{2}$ second at f/22 on ISO 200 slide film. The right image of Jupiter emerging was taken just before the moon set so the telecompressor was not used in order to minimize the required exposure time. The exposure is $\frac{1}{2}$ second at f/11. This image is enlarged about 2.2×; twice as much as the other 3. Respective times for the photos are 9:37, 9:42, 9:44, and 10:26 pm mountain daylight time. Bottom: The 3 July, 1989 occultation of the 5th magnitude star SAO 187255 by Saturn was captured with a 20 cm f/10 SCT, a 2× Barlow lens, and a C-mount $\frac{1}{2}$ inch format monochrome CCD video camera. From left to right, the images show the event at 5:54, 6:02, 6:14, 6:47, 8:35, 9:12, 9:25, and 9:45 Universal Time (U.T.). The 6:14 and 9:12 images show the star shining through Cassini's division in Saturn's rings.

makes the event easy to observe with the unaided eye as well as reducing requirements for equipment you need to photograph it. Here, you can get a nice shot with a camera and a modest telephoto lens. You can get an even better occultation picture with a good small telescope. If you have a tracking mount for your camera or telescope, you can even photograph earth-shine on the crescent moon, as shown in Figure 9.2.

The same equipment can be used to photograph an occultation by an old crescent moon, but in this case, the dark part of the moon will be on the side from which the occulted object emerges.

If the moon is mostly illuminated when it occults an object, you may need at least a small telescope to adequately photograph it. In general, a focal length of at least 1000 mm is required to get a good film image of a planet when it is partially covered by an illuminated part of the moon. A focal length of 2000 mm or even more would be preferable, since this at least has the potential to capture a little detail on a planet such as Jupiter or Saturn.

You can get even larger images of the moon progressively covering a major planet by using the same techniques of eyepiece projection or afocal photography that you would use for ordinary planetary photos. A really long focal length will not allow you to get the entire moon in your picture, but it can definitely provide some impressive close up occultation images!

9.3.2 Occultations by Other Solar System Objects

Occultations by objects other than the moon also occur. Occasionally, an asteroid or major planet will cover a star. These events are not always as dramatic as occultations by the moon, but they are just as useful for scientific purposes. As mentioned in chapter 2, an occultation of a star by an asteroid provides an excellent way to discover the size and profile of the asteroid. The same can be true in the case of a comet if its nucleus occults a sufficiently bright star.

Occultations by some major planets have revealed remarkably detailed data on their atmosphere and ring structure. When Saturn occulted a 5th magnitude star in 1989, over 1,000 photometric events were recorded by some observers, and this data provided valuable information about the structure of Saturn's rings. Likewise, some spacecraft NASA sent to other planets used occultations of stars to get ring data.

You can observe some occultations of stars by asteroids or major planets with a small telescope, but getting good photos or electronic images of these events can require at least a moderate sized telescope. In general, you do not need as much aperture for electronic imaging as you do for film because the small pixel size of most efficient modern sensors (and the correspondingly shorter required focal length) permits a shorter exposure for a given amount of imaged detail. You may be able to get by with 10 or 20 cm of aperture for photographing an occultation by a major planet, but an aperture of at least 30 cm may be required to adequately photograph a typical occultation of a dim star by an asteroid.

If you have only a small telescope, you can always observe and document (draw or time) this type of occultation rather than taking photos. Images are not required for observations of asteroid occultations; it's the timing data that matters most for scientific work. As is the case for other events which are too dim and brief to adequately photograph with a small telescope, you can always make drawings of your observations.

Other events worth noting include mutual eclipses by the moons of Jupiter or Saturn. While not an occultation, rare collisions between celestial objects such as the 1994 collision of comet Shoemaker–Levy 9 with Jupiter certainly are dramatic through a telescope!

9.4 Lunar Eclipses

Lunar eclipses occur when the moon moves through the earth's shadow. The eclipse can be penumbral, partial, or total, depending on how near the center of the earth's shadow the moon gets as it passes by. This variation is possible because the orbit of the moon is inclined about 5 degrees from the plane of the earth's orbit.

If there were no inclination of the moon's orbit, there would be a total lunar eclipse every 29 days or so. Due to its orbital inclination, the moon passes well to the north or south of the earth's shadow most of the time. Only a few times a year does the moon cross the earth's orbital plane at or near the time it is in line with the earth and sun, making some sort of eclipse possible. The points at which the moon crosses the earth's orbital plane are called nodes. These nodes are referred to as ascending or descending, depending on whether the moon is moving to the northern or southern side of the earth's orbital plane.

The sun is an extended object, so shadows cast into its light do not have perfectly sharp edges. Instead, the shadow is composed of two parts. The central umbra is

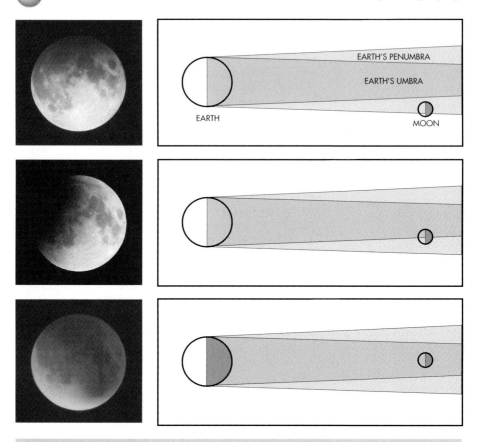

Figure 9.4. A lunar eclipse is observable from any location on earth in which the moon is above the horizon and local weather is sufficiently clear. Top left: Appearance of the moon during a penumbral lunar eclipse. Photographed at 1/125 second on ISO 200 slide film with a 10.2 cm f/15 refractor. Top right: Circumstances for a penumbral lunar eclipse. Middle left: Appearance of the moon during a partial umbral lunar eclipse. 1/30 second exposure on ISO 200 slide film through a 9 cm f/11 Maksutov–Cassegrain telescope. Middle right: Circumstances for a partial umbral lunar eclipse. Bottom left: Appearance of the moon during the 6 September 1979 total lunar eclipse. 35 second exposure on ISO 200 slide film through a 9 cm f/11 Maksutov–Cassegrain telescope which was on an equatorial mount. Bottom right: Circumstances for a total lunar eclipse.

an area of complete shadow where the sun is totally eclipsed. The surrounding penumbra is an area of partial shadow. From the penumbra, the sun appears partially eclipsed, with the area of least eclipse comprising the penumbra's outer edge.

In a penumbral eclipse, the moon enters an area which is only partially shadowed by the earth. This type of eclipse is not always obvious visually, but it can often be detected on a properly exposed slide, partly due to the relatively high contrast of slide film.

A very slight penumbral eclipse can be of relatively short duration, but most last a few hours due to the large diameter of the earth's penumbra. A deep penumbral eclipse can make one pole of the moon appear a bit dusky for anywhere between several minutes to an hour or so. The dusky appearance is due to one pole of the moon being relatively close to the earth's umbra.

In a partial umbral eclipse (also called a "partial lunar eclipse"), the edge of the earth's round umbral shadow actually covers part of the moon. The appearance of a partial eclipse can vary between a mere darkening of one edge of the moon to a situation in which a mere "fingernail" of the moon is directly illuminated.

When a large portion of the moon is within the earth's umbra, it is not unusual to see color on the shadowed part of its surface. The color, which can range from red or orange to a washed out tan or butterscotch hue, is caused by refraction of sunlight through the earth's atmosphere. This is the only significant illumination inside the earth's umbral shadow. The earth's umbra is usually darkest toward the center because more refraction is required to provide illumination. Additional refraction means that the light must pass through more of the atmosphere, where still more of it is absorbed or blocked.

The warm hue within the umbra is caused by the same factors that cause sunlight here on earth to have a warm color near sunrise or sunset. Namely, shorter (bluer) wavelengths of light tend to be scattered or absorbed by the atmosphere, while longer wavelengths such as yellow or red pass through it to a greater degree. If the earth had no atmosphere, the totally shadowed part of the moon would instead appear to be about as dark as the surrounding night sky, making it more or less invisible.

9.4.1 Total Lunar Eclipses

The most dramatic of lunar eclipses is a total one. During a total lunar eclipse, the entire moon is within the earth's umbra. It is then that colors within the umbra can be observed most easily on the moon. In addition, a total lunar eclipse can cause the character of the entire night sky to change if you are away from city lights. As the eclipse begins, light from the full moon washes out the sky, but as more of the moon enters the earth's umbra, the entire sky darkens and celestial features such as the Milky Way become more obvious.

A total lunar eclipse begins with a penumbral eclipse, then progresses to an umbral eclipse. The transition between each stage of the eclipse is called a "contact". First contact is when the edge (or limb) of the moon first enters the earth's penumbra. Second contact is when the edge of the moon reaches the umbra. Third contact is when the entire moon finishes entering the earth's umbra and the total phase of the eclipse begins. Fourth, fifth, and sixth contacts correspond to when the moon moves out of the umbra and penumbra. The beginning and ending partial phases can each last an hour or more, as can totality. Such a slow pace helps make observing and photographing a lunar eclipse an enjoyable and relaxing experience.

Colors within the earth's umbra are most obvious during the total phase of the eclipse, but some color can be detected even when the eclipse is not total. The part of the moon within the earth's umbra is typically visible under moderate magnification during almost any stage of a partial eclipse. By the time about 50 percent of the moon is within the umbra, the shadowed part of the moon can become relatively easy to see with the naked eye.

As the total phase of the eclipse approaches, less and less of the moon is in direct sunlight, and penumbral darkening near the umbral boundary reduces the brilliance of the remaining slice of sunlight along the lunar limb. During this time, colors on the part of the moon within the umbra gradually become more obvious. The transition to totality often appears to be gradual; so gradual that it can be difficult to time the beginning or end of totality down to the second.

Brightness of the total phase of a lunar eclipse can vary widely, depending on atmospheric conditions here on earth and how close the moon gets to the center of the umbra. When the earth's atmosphere is relatively clear, the totally eclipsed moon may be so bright that you can photograph it with a camera on a fixed tripod. If there is a lot of obscuring material in the atmosphere (as can happen after a big volcanic eruption) a central total lunar eclipse can be so dim that you can't even see the moon with the unaided eye! In such cases, an exposure of more than one minute at f/2.8 may be required to photograph the eclipse on ISO 400 film.

In addition to interesting colors, total lunar eclipses can have other unique attributes. For instance, the total lunar eclipse of 6 September 1979 revealed a curious and relatively well defined dark feature in the earth's umbra. (This feature is visible toward the lower right of the total lunar eclipse photo in Figure 9.4.) In most subsequent total lunar eclipses, the earth's umbra appeared to be more homogeneous.

You can photograph a lunar eclipse with either a camera lens or a telescope. Even a modest camera lens can provide interesting pictures of these events, but a telescope or extreme telephoto lens will allow you to better record the edge of the earth's shadow as it covers features on the moon. The earth's penumbra causes significant darkening near the umbral boundary, and the earth's atmosphere causes some sunlight to be refracted into the umbra. The combined effect of these factors tends to reduce the apparent sharpness of the umbral boundary. This in turn causes pictures of varying exposures to show the umbra in slightly differing apparent positions, and the effect can be so pronounced that the moon may appear totally eclipsed in a dim photograph which is taken several minutes before totality. In order to prevent this from happening in your pictures, it is usually best to gradually increase the exposure as the eclipse deepens, then decrease the exposure again as the moon begins to exit the umbra.

Figure 9.5. A dim lunar eclipse combined with a dark site can make for dramatic images. This 5 minute exposure of the exceptionally dark total lunar eclipse of 30 December 1982 was taken with a 105 mm f/2.5 lens on Ektachrome 200 film. This eclipse was so dim that the moon was hard to locate with the unaided eye. Blurring of the moon is caused by its motion relative to the background stars. A shorter exposure (as would be possible with a faster film or during a brighter eclipse) would result in less visible blurring. This cropped image is enlarged about 7×. The star cluster toward the upper right is M35.

About 10 or 20 minutes before the partial umbral phase of the eclipse begins, I usually start out with about half an f/stop more exposure than I'd normally use for the full moon. As the eclipse approaches about 30 percent, I usually increase the exposure one f/stop by going to a slower shutter speed. As the eclipse nears 80 percent, I increase the exposure another stop, then gradually increase the exposure from one to three more f/stops as the eclipse nears totality. A minute or two before the eclipse becomes total, I take a series of increasingly long exposures until I reach the proper range of exposure for totality. This provides a series of images to select from which should give you a smooth transition from the partial to the total eclipse phase.

If your emphasis is instead to accurately document the progress of the umbra as it covers the moon, it is best to use the same exposure for all photos of the partial phases. Here, the best exposure is about two or three f/stops more than what you would normally use to photograph the full moon.

During any part of the partial phase of the eclipse, you can photograph the part of the moon within the umbra simply by using an appropriately long exposure. Pictures taken during about a 50 percent eclipse can be quite interesting. If you have the right equipment for digital image processing or fancy darkroom work, you can later combine such an image with one having the correct exposure for the directly illuminated portion of the moon. The final result can be more reminiscent of what this stage of the eclipse looks like visually.

If the lunar eclipse is tracked during totality, you can photograph it in front of background stars. If the background stars are tracked and you take one shot every hour or so, you can get a sequence picture which shows the boundary of the earth's umbra projected on a series of lunar images.

If you want to get a more exact image of the earth's umbral dimensions, you will need to dispense with using the background stars as a fixed reference and instead use an offset registration technique to compensate for umbral motion relative to the stars which results from the earth's motion in its orbit. Since the earth moves a measurable distance (and therefore a measurable angle) in its orbit around the sun during the hours of the lunar eclipse, the cumulative change in its angle to the sun (plus the change in your relative position due to the earth's rotation on its axis) will cause the umbra to move a fraction of a degree in relation to the background stars.

If your camera is only on a fixed tripod, you can record partial phases of a partial or total lunar eclipse in separate pictures or a single sequence picture. If you use a wide angle lens at a very slow aperture with relatively slow film, you can even record a total lunar eclipse as an interesting streak which starts out bright, dims to a pale red or orange, then brightens again.

9.5 Solar Eclipses

Solar eclipses occur when the moon moves between the earth and the sun. Such eclipses can be partial, annular, or total. As with lunar eclipses, a solar eclipse can only occur when the moon crosses the earth's orbital plane at or near a time it is between the earth and sun. The type of solar eclipse that occurs is dependent on the distance between the earth and the moon at the time, the terrestrial location from which the eclipse is observed, and how close the moon is to crossing the earth's orbital plane.

When the moon crosses the earth's orbital plane a day or so from the time it is in line with the sun, the eclipse will only be partial. The degree of partial eclipse will determine what effects it has on the local environment. In very strong partial eclipses, the effect will also be determined by the distance to the moon, since this will affect its apparent size in relation to the sun.

If the sun is only about 30 percent eclipsed, little evidence of it will be obvious to most people unless they look at the sun (something that should only be done with proper filtration). Dimming of the ambient light becomes easy to notice when an eclipse reaches about 70 percent, and it is pronounced at 90 percent or so. During a strong eclipse, local conditions can become so dim that street lights (at least those having dusk to dawn sensors) will come on. This can also be true of a more moderate eclipse if it occurs on a cloudy day or when the sun is relatively near the horizon.

If the moon crosses the earth's orbital plane close to the time of the eclipse, the moon can move directly between the sun and part of the earth's surface. If you are on the right part of the earth's surface, the moon will appear to move directly in front of the sun. When this occurs, the eclipse can be either annular or total, depending on the distance between the earth and the moon. On rare occasions, an eclipse can be a very thin

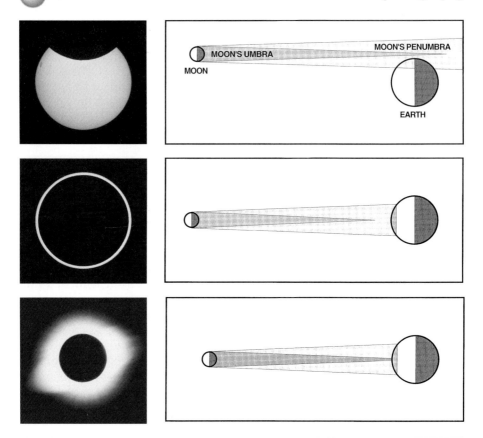

Figure 9.6. Top left: Appearance of the sun during a partial solar eclipse. Photographed with an 8.9 cm f/16 Maksutov–Cassegrain telescope having a front solar filter. Top right: Circumstances for a partial solar eclipse. Middle left: Appearance of the sun during an annular solar eclipse. Simulated in Adobe Photoshop. Middle right: Circumstances for an annular solar eclipse. Bottom left: Appearance of the solar corona during the 24 October 1995 total solar eclipse. Exposed for one second on ISO 64 film through a 300 mm lens and 2× Barlow lens. A solar filter is not required during the brief seconds or minutes of totality. Bottom right: Circumstances for a total solar eclipse. Total and annular eclipses are only visible from a very narrow path on the earth's surface, but the related partial solar eclipse may be observed from up to about a third of the sunlit part of the earth.

annular one at the beginning and end of its path and total near the middle. Such an eclipse will occur in the year 2005.

The sun is far larger than the moon, so the lunar umbra (the area of complete shadow behind the moon) tapers down to a point. The moon's distance from the earth varies as it orbits the earth in an elliptical path. About half the time, the angular size of the moon is smaller than that of the sun. If a central eclipse occurs when the moon is relatively distant, the eclipse will be annular because the tip of the lunar umbra does not reach the earth. During such an eclipse, the sun appears as a relatively thin ring around the backlit moon. Local ambient light is noticeably subdued during an annular eclipse, and in cases where the annular phase lasts a couple of minutes or less, street lights having light sensors will usually come on.

Photos of partial and annular solar eclipses are typically taken with a front solar filter and the photographic technique is the same as that for taking any other solar picture. A useful image of a partial or annular eclipse can be acquired on the 35 mm format with a focal length of 200 mm to 2000 mm, but 800 mm or longer is best because it will provide enough resolution to show larger sunspots (if there are any) relatively well. In addition, a long focal length can reveal any obvious roughness on the silhouetted lunar limb.

A multiple exposure sequence image can be taken with a normal or wide angle camera lens and a proper solar filter. If you want to capture the entire duration of an eclipse in one picture from a fixed tripod, your lens will have to be wide enough to cover the solar position at both the beginning and end of the event. Since the sun moves across the sky at about 15 degrees per hour, the angle of your lens will typically have to be more than the duration of the eclipse in hours, multiplied by 15 degrees. For example, if the entire eclipse lasts 2 hours and 20 minutes, the sun will move across up to 35 degrees of sky between the beginning and the end of the event. If you use a 35 mm camera, a normal 50 mm lens would just barely cover it, since such a lens has a horizontal coverage of about 40 degrees.

For partial eclipses of up to about 95 percent, you can use the same filtered exposure as that required for the uneclipsed sun. If the eclipse is more extreme or if it is annular, the exposure should be increased one or two f/stops in order to compensate for solar limb darkening.

If you get a new solar filter specifically to photograph an eclipse, it is advisable to take test pictures through the filter because there can be a wide variation in the density of different commercial solar filters.

During very strong partial eclipses, photos have been taken which show prominences and even some corona in areas where the solar limb is just barely covered by the moon. Such pictures are taken without solar filters, but you obviously cannot look in your camera finder when the solar filter is removed from your optics. To take such a picture, the filter should be removed only while the picture is taken. Failure to replace it right away could cause damage to your camera or internal baffles in your telescope.

It is not safe to look directly at any annular or partial eclipse without proper filtration (nor is it safe to look in your SLR camera when your optics are not properly filtered), no matter how extreme the partial eclipse may be.

It is always unsafe to look at the sun without proper filtration, but special caution is advised for eclipses because this is the time people may be most tempted to look at the sun. A partial solar eclipse is no more dangerous to look at than the uneclipsed sun for a given amount of time, but it can do more damage because the total amount of glare from the sun can be reduced to such an extent that you may not feel any pain at the time you look at it. Glare around the uneclipsed sun makes it unpleasant to look in its direction, but a strong eclipse may have so little glare that you could stare at it for several seconds without feeling any pain right away. Meanwhile, your retina gets fried and may hemorrhage, resulting in blindness (maybe even permanent blindness) anywhere from a few seconds to a few hours later. Only a rare total eclipse (covered below) can be observed without filtration.

In order to be safe, a visual solar filter must block all but about 1/100,000 of the sun's light. Such a filter should be free of defects such as a thin coating, or holes or scratches in the coating which can make the image too bright. Through a safe solar filter, the sun should look relatively dim. If a commercial solar provides a solar image which looks harsh and bright, it may be defective.

It is not safe to use methods providing less filtration than a safe solar filter. Dangerous solar observation methods include looking at a reflection of the sun in water or a pane of glass. The surface of water reflects about 3 percent of the sun's light; some 3,000 times more than what is safe to look at. Glass reflects even more; about 4 percent per surface, so neither of these reflective viewing methods are safe.

If you find yourself at an eclipse without any filter but you want to observe it, punch a two millimeter or smaller hole in a piece of cardboard and hold it between the sun and a projection surface such as a piece of white paper, then look at the solar image which is projected onto the projection surface. A projection distance of half a meter or more will work relatively well. Another way of looking at a solar eclipse is to use a small mirror to cast its reflection on an object a few meters away. If you lack any of the above items, look around for a leafy tree. Small gaps between leaves often cause multiple projections of an eclipse to be visible under a tree.

9.5.1 Total Solar Eclipses

A total solar eclipse is the ultimate astronomical event. In mere moments, the appearance of everything seems to change as you are engulfed by the shadow of the moon. Wild and domestic animals alike are fooled in to thinking sunset is near as the eclipse progresses. Roosters crow and other birds fly back to their nests. The experience of witnessing a total solar eclipse is so grand that we humans can be affected as well; some of us in ways that last for a lifetime.

If the moon is relatively close to earth, a central solar eclipse will be total, but only within a small area where the conical lunar umbra intersects the earth's surface. Just outside the umbra, the eclipse is only partial. As the moon passes between the earth and the sun, the lunar umbra usually moves across the earth along a predominantly west to east path. A total solar eclipse occurs at sunrise on the western end of the path, near noon at the center of the path, and at sunset on the eastern end of the path. There are rare exceptions when the umbra just skims polar latitudes of the earth. At such times, the total eclipse may occur at either sunrise or sunset at both ends of the shortened eclipse path.

The partial phases of a solar eclipse can last well over an hour, but the total phase, or totality, lasts only a few seconds or minutes. The longest possible duration of totality is only about $7\frac{1}{2}$ minutes, and most total eclipses are far shorter, with 2–5 minutes of totality being more common.

The longest duration of totality usually occurs on the part of the eclipse path where maximum eclipse occurs near noon. This is because the curvature of the earth places the observer a few thousand kilometers closer to the moon (making it appear larger) and the earth's rotation carries the observer in predominantly the same direction that the moon's shadow moves, thereby slowing the shadow's ground speed. During totality, the bright solar photosphere is completely covered and the fainter pearly white corona can be seen in all its splendor.

A total solar eclipse begins with a partial eclipse, and as with a lunar or annular eclipse, the transition between each stage is called a "contact". First contact is when the edge (or limb) of the moon first begins to cover the sun. Second contact is when the total eclipse begins. Third contact is when totality ends, and fourth contact is when the moon no longer covers any part of the sun.

Many interesting things happen during a total solar eclipse, and the short duration of totality makes such an eclipse one of the most challenging photographic subjects in nature. The solar corona is what comes to mind when most think of a total solar eclipse, but there are many other fascinating aspects of the event.

The lunar umbra can easily have a diameter of 100 kilometers or more at the earth's surface, and its approaching darkness may actually be visible in the western sky up to several minutes before the total phase of the eclipse begins. At first, it may be very subtle, but by a couple of minutes before totality it can (but does not always) cause part of the sky to take on a dark blue color similar to what you would see under a late afternoon thunderstorm.

If the eclipse occurs in the morning, the umbra can begin to darken the sky directly overhead up to a minute or so before totality. If the eclipse occurs in the afternoon, the umbra will typically approach from an azimuth at or near that of the sun, darkening the sky below the eclipse. As totality draws near, it may be possible to see the umbra darken surrounding terrestrial features such as distant cloud formations or mountain ranges.

The ambient light level can drop faster than three f/stops per minute just before the total phase of an eclipse, and this is quite exciting! At the beginning of totality, the umbra engulfs eclipse observers and everything around them. It is then that intensity of the ambient light drops the fastest. Throughout totality, ambient light intensity changes relatively little, though there is a gradual shift in brightness between the lower eastern and the lower western sky as the umbra moves in a generally eastward direction.

During totality, an exposure as long as one or two seconds at f/4 on ISO 100 film may be required to get a good picture of the eerily dim surroundings. It is not completely dark, but it is usually dim enough to make reading difficult. If you expose a terrestrial picture at about two f/stops less than the incident light reading, it will tend to provide better saturation in the warm

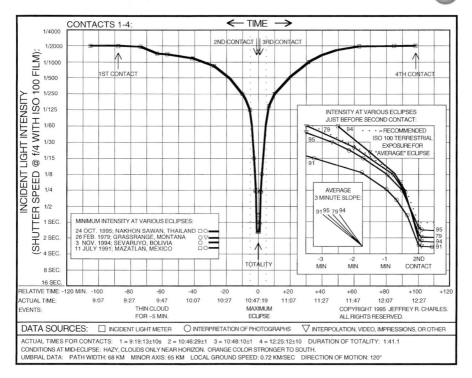

Figure 9.7. This graph of incident light intensity at the 24 October 1995 total solar eclipse was generated from light meter readings and photographic data. The rapid change in ambient light intensity near the time of totality can make it challenging to get good photos of terrestrial effects. This graph can be used as a sort of exposure guide for photographing terrestrial scenes during various stages of a total solar eclipse. The inset at lower right shows the drop in light intensity within three minutes of totality at a few other total eclipses.

sunset colors which may appear around the horizon. The sunset colors usually appear most realistic in pictures shot on color negative film, because its contrast is lower than that of slide film.

9.5.1.1 The Solar Corona

The solar corona gradually becomes visible over a period of a few seconds or minutes which immediately precede totality, but it is not safe to look at the sun without a proper filter until totality actually occurs. You can photograph the inner corona before and after totality if you are careful. To do so, just remove your solar filter, take a picture at or near your fastest shutter speed, then immediately replace your solar filter; but DO NOT try to look in your telescope or camera finder while the solar filter is off until the eclipse is total. In using this unfiltered method, I have photographed the inner corona up to about two minutes before and after totality. If atmospheric conditions are really clear, I would not be surprised if one could capture some of the inner corona at least 5 minutes before or after totality.

Just before the eclipse becomes total, the moon covers all of the bright solar photosphere except a narrow sliver of a crescent. At this point, it is not unusual for mountains on the moon to cause ends of the shortening crescent to break off into beadlike dots (called "Baily's beads") and fade away. When only a few degrees of the crescent remain, glare from the tiny amount of remaining solar photosphere is so slight that it no longer brightens much of the surrounding sky. Instead, the vanishing glare around the sunlight shrinks until it is limited to an angular area smaller than the moon, allowing the solar corona to become increasingly obvious. This shrinking area of bright sunlight and glare set against the corona is called the "diamond ring" effect.

As totality begins, the last remnant of the solar photosphere often breaks up into beads (called Baily's beads) as it is quickly covered by the moon. The solar eclipse is then total and one can safely look directly at it and photograph it without a solar filter.

With the last of the solar photosphere covered, the solar chromosphere becomes more obvious, appearing to be a rapidly shrinking arc of pink light. In a few seconds, it is gone, but many areas of pink light may remain. These are prominences, and those large enough to easily see with the naked eye are typically larger than the earth.

As your eyes adjust to the dusky light of totality, the full extent of the solar corona becomes easy to see, but it is best appreciated if you look at it for several tens of seconds rather than just taking a quick glance. The corona is fully visible at the moment of totality, but it does not suddenly "flash" into view as it appears to do on some videos. The full brightness of the corona was there before totality; it was simply obscured by glare from the brighter solar photosphere until totality.

The corona is rich with detail. In a clear sky, fanlike polar streamers usually appear to extend out to about a solar diameter in most directions. It is not unusual for the corona to have some equatorial streamers which extend out to two solar diameters or more; I have even seen streamers longer than five diameters. The appearance of the corona is different at each eclipse. Near sunspot maximum, the corona may look nearly round or have long streamers which go out in many directions. At other times, the longest streamers will tend to extend only toward the east and west.

If thin clouds are present, the corona may only appear to extend out to half a solar diameter or even less. Thin clouds can make the corona look a lot less impressive, though they usually do not significantly affect your ability to observe and photograph prominences.

Figure 9.8. Upper left: The solar chromosphere and a small prominence are visible in this enlargement from a 1/500 exposure at f/11 on Kodachrome 64 professional film. This photo of the 3 November 1994 total solar eclipse was taken with an unfiltered Vernonscope 9.4 cm f/7 refractor and 1.6× Barlow lens. The image is enlarged about 4.5×. Upper right: Clear skies allowed this three second exposure with the same unfiltered telescope and film to clearly reveal earthshine on the moon during totality. A Barlow lens was not used for this picture, so the telescope was working at its original focal length of 640 mm. This figure is a 4× enlargement. Lower left: This $\frac{1}{2}$ second exposure captured particularly striking polar streamers in the corona. Lower right: The diamond ring marks the end of totality at the 24 October 1995 total solar eclipse. This is a one second exposure on Kodachrome 64 film with an unfiltered 300 mm f/4.5 ED Nikkor lens and a Versacorp VersaScope™ adapter (an attachment for the Versacorp DiaGuider) which includes an integral 2× Barlow lens.

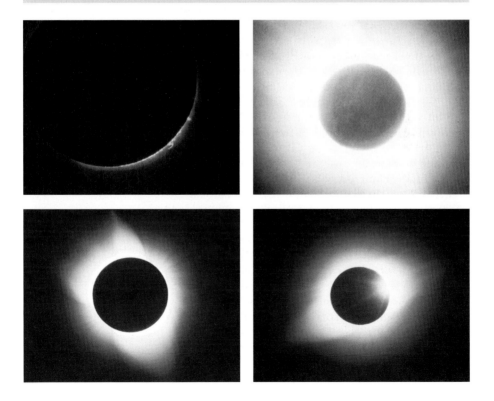

As totality draws to a close, the solar chromosphere becomes visible on the western side of the moon, then a repeat of the diamond ring effect (on the opposite side of the moon from the first one) heralds the end of totality. At this point, it is no longer safe to look at the sun without proper filtration.

Some people observe Baily's beads and the diamond ring, while others do not observe it for safety reasons. Those that do observe it usually do so with the naked eye and look away from the eclipse no more than 5 seconds after the first part of the photosphere becomes visible. You definitely would not want to observe the diamond ring (or any other part of the partial phase of an eclipse) through an unfiltered binocular or telescope. Such optics can magnify the intensity of the emerging crescent of solar photosphere to such an extent that blindness could result in a fraction of a second.

After totality, the moon gradually uncovers the entire sun. Normal daylight gradually returns, but to those who have seen a total solar eclipse, the sun is no longer a mere ball of light that shines during the day. Instead, some regard its illusive corona as one of the most beautiful of celestial objects. The moon covered the sun, yet the sun was revealed in a way you've never seen it before. It's then that you ask: "When is the next eclipse?"

A total solar eclipse is truly exciting, and it is only natural to want to take pictures of it. The experience can be overwhelming, and all the excitement can make it difficult to operate your cameras properly unless you have practiced your procedures.

A lot of things happen at once during the transition to totality, and you may need to use several cameras if you want to capture the full scope of the event; however, it is usually best not to overdo it. If you try to use too many cameras, you run the risk of not having time to *look* at the eclipse, and if you don't look at the eclipse, you won't know how your pictures should look when you print them!

It is best to keep things simple for your first eclipse in particular. Many people may only plan to see one eclipse, but total solar eclipses have a way of being addictive. There is nothing quite like being inside the shadow of the moon. It's just a matter of "getting in the shade", but what a special kind if shade it is! Once you've seen one total eclipse, you too may be bitten by the eclipse bug and start going to every one you can.

Repeated practice of eclipse photography procedures in concert with a timer will allow you to become familiar with using your photographic equipment under pressure, and if you practice your procedures while a video or computer simulation of an eclipse is playing, you will better appreciate the necessity for being organized. The eclipse won't wait for you, and if totality ends before you get the pictures you want, you won't get a second chance. All you can do is wait for the next one.

The astrophotography exposure guide in Appendix A of this book provides useful exposure data for solar eclipses and other events. After having seen an eclipse under perfectly clear skies, my own preference has been to emphasize long exposures of about a second or two at f/11 on ISO 100 film. These best reveal the outer corona and are most reminiscent of what a total solar eclipse actually looks like when seen with the naked eye. It is usually best to shoot series of exposures which range from the equivalent of about 4 seconds at f/11 to at least 1/250 second at the same f/ratio on ISO 100 film. The short exposures are for prominences and the extreme inner corona.

Different exposures of the diamond ring will change the amount of corona in your image. A long exposure will show a lot of corona and make the diamond ring effect look more subtle. An exposure of 1/15 second or so at f/11 on ISO 100 film will make the corona look like a thinner "ring" which has a more pronounced "diamond".

I have found slide film to be best for prominences and print film to be best for the corona, though slides have proven adequate for every aspect if I composite multiple corona images or do much in the way of dodging. My favorite slide film for eclipses is Kodachrome 64 professional. I don't like regular Kodachrome as much because it tends to have a more pronounced turquoise or greenish cast in moderately dim parts of the picture. In reality, the entire corona looks about the same pearly white color, so a good black and white film can capture it just as well as color films. The pink color of prominences and the grayish blue color of surrounding sky can always be added back into the image (either digitally or in the darkroom or by manual retouching) when a final print is made.

An adequate picture of the solar corona can be obtained from a fixed tripod with a lens as short as 200 mm, but a focal length of 600–800 mm is usually best. A still longer focal length can better allow you to photograph prominences, but too long a focal length (more than about 1600 mm for the 35 mm format) will keep you from getting even the entire inner corona in one picture.

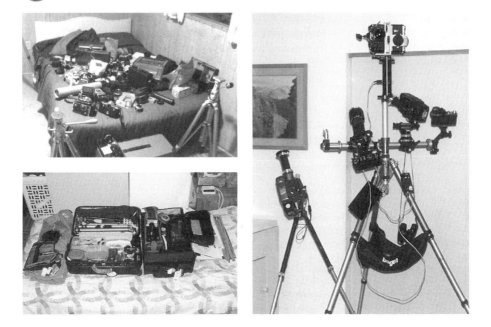

Figure 9.9. The degree of success you have in eclipse photography may be dependent on how well you organize your equipment and procedures. Upper left: The hodgepodge of second hand equipment I used at my first total solar eclipse on 26 February 1979 covered an entire double bed and a good part of the floor. This picture shows that the center of the bed was literally sagging under all the weight! Nearly every camera was on a separate tripod. This required a lot of luggage space and setup time. Lower left: Having learned my lesson about trying to bring the equivalent of an entire camera store to total solar eclipses, I developed a portable system for the 24 October 1995 eclipse which could be transported with my clothing in two pieces of carry-on luggage. As a result, I got better pictures. Right: Dubbed the "Christmas tree" by another observer, this well tuned setup can be transported in the two previously shown pieces of carry-on luggage. It includes five still cameras, two video cameras, and an automated indexing rotary camera platform for shooting 360 degree panoramas. The modified Bogen cross-bar allows several cameras to be mounted on a single tripod, so fewer tripods (and fewer pieces of baggage) are required. I did not bring a "real" telescope or sidereal drive to the 1995 eclipse because I wanted to simplify things and my 300 mm lens combined with a VersaScope™ teleconverter adapter and a DiaGuider™ were adequate for the outer corona shots I was after.

9.5.1.2 The Lunar Umbra

Photographing the lunar umbra is not all that popular among most observers, but it is my favorite subject at eclipses because it is what affects my immediate surroundings the most. Capturing the full scope of its dramatic effect can require a fisheye lens or even specialized panoramic equipment.

The umbral boundary is often visible as a diffuse transition between the sunlit sky outside the umbra and the dark gray-blue color within. Depending on local conditions, the umbra may be visible up to several minutes before and after totality, but it is usually the most dramatic within a minute or two of totality. This conclusion is consistent with the preliminary results of experiments I have performed which show that the umbral boundary is most obviously projected on the clear sky at an altitude of about 22 kilometers. These results are not surprising, since light scattering material such as ozone and volcanic dust are relatively plentiful at this altitude.

If low level haze is present but the sky is relatively free of clouds, it is not unusual to also see the umbra as a grayish area right on the horizon as it shadows the haze layer. When thin cirrus clouds are present, the umbral boundary is typically most visible on them. In fact, high thin clouds can reveal the umbral boundary so well that its round shape is obvious in wide angle pictures and visual observations made with fisheye "door peeper" lenses.

Immersive 360 degree panoramas are best for capturing the true feel of what a total solar eclipse does to your surroundings. Extremely wide angle optics are particularly useful for capturing such a panorama in one shot. When used in a vertical orientation, a wide angle reflector of sufficient curvature can capture an entire 360 degree panorama of the sky and a variety of terrestrial features in a single picture. The original image is circular, but can be converted to a straight panorama with programs such as Adobe Photoshop, then to computer readable viewing formats such as Quick Time VR. Straight panoramas are desirable as the end result because they provide a better feel for some subjects than circular images.

Higher resolution one shot panoramas can be obtained with wide angle reflectors if a large film format (or high resolution electronic sensor) is used, but if the benefits of the large format are to be fully realized, special optics are required to correct for field curvature and various aberrations inherent in simple reflective panoramic imaging systems.

Even higher resolution can be achieved on a given format by assembling a series of wide angle pictures into a 360 degree panorama, but such pictures need to be taken quickly during the precious seconds or minutes of totality. To facilitate this, I invented and built a motorized indexing rotary panoramic camera platform for the 11 July 1991 total solar eclipse. This prototype Versarama™ platform enables me to shoot each 360 degree panorama in less than 6 seconds by just pushing a couple

Figure 9.10. Upper left: The wide 130 degree horizontal coverage of a 16 mm fisheye lens captures the rounded edge of the retreating lunar umbra about 15 seconds after the end of totality at the 26 February, 1979 total solar eclipse. I recorded the exposure as 1/15 second at f/4 on ISO 100 film but suspect that I may have used a slower shutter speed. Upper right: An upward pointing fisheye lens captures the approaching lunar umbra about a minute before totality at the 25 October 1995 total solar eclipse. The exposure is $\frac{1}{4}$ second at f/9.5 on ISO 64 slide film. Bottom: High clouds may not be good for observation of the corona, but they can provide a dramatic projection screen for the lunar umbra! This 360 degree panorama of the lunar umbra approaching Mazatlan, Mexico was captured a mere 20 seconds before totality at the 11 July 1991 total solar eclipse with the use of the prototype Versarama™ motorized indexing rotary camera platform shown in Figure 9.9. Symbols for various directions are shown under the panorama.

of buttons. I later added a circuit to provide automatic operation of the camera and platform at later eclipses. The 360 degree panorama shown in this section is made up of four 1/15 second exposures which were taken with the platform and a 16 mm fisheye lens on ISO 125 color negative film. One of the four pictures is repeated at each end of the panorama to provide overlap.

Even though I developed cutting edge wide angle optics (patent pending) which capture unobstructed circular 360 degree panoramas in a single shot, I still emphasize use of my rotary platform to shoot eclipse panoramas as a sequence of separate pictures because it provides higher resolution on a given film format. A high resolution panorama permits better enlargement of areas which are most interesting.

Chapter 10

After You Take Your Astrophotos

Taking astrophotos is the first step toward getting the final images you want. If you shoot negatives, you will either need to print or scan them in order to adequately see your results. If you shoot slides, you will be able to see your results in your original processed film, but printing them in the way you want may be more difficult. You can do a lot with an image after the film is developed, but this is no substitute for a good original image. The better your original image, the better your final result.

10.1 Film Developing

Astrophotos are unconventional in many ways, so some astrophotographers develop their own film. Black and white film is relatively easy to develop because the process is simple, usually consisting only of developer, stop bath, fixer, and a long rinse. The process is tolerant of several degrees variation in temperature, though too much variation (particularly when the change is sudden) can cause problems.

Color film developing processes are more particular about temperature and time. For color negatives, too cold a developer temperature (even one degree in some cases) or too little development time can result in thin negatives which have poor color saturation and contrast. Too warm a developer temperature can result in too much base density. Too much development time (sometimes as little as 5 percent too much) can cause unrealistically high contrast and color saturation that makes your pictures look almost like cartoons. Such overdevelopment can bring out brilliant colors in nebulae, but it also tends to increase graininess and limit the film's dynamic range.

Some commercial film labs can adequately process and even print your astrophotos, but they can only do so if they know what your requirements are. A few precautions are in order if you use a commercial lab. One is to preview their work by asking to see samples of negatives or prints they have processed. If they show you negatives from one of their current orders and they look all right, then the lab may be worth trying. If instead the negatives are scratched or have spots or finger prints on them, you may want to look for another lab.

Another precaution is to take a picture of something showing your name and address at the beginning of each roll; or at least shoot a picture of a subject bright enough to define the edges of the format. This will provide an image you can use as an index when you cut your film. Deep sky astrophotos can resemble unexposed frames at first glance, so an uninformed photo technician may think film containing astrophotos has not been exposed at all. Therefore, it is best not to let a photo lab cut your film; just tell them not to cut (or mount in the case of slides) your film. The last thing you need is for your film to be cut right through the center of one of your best pictures! Ideally, the lab you use will be one in which you can talk directly with the technician. Some one hour photo labs may facilitate this.

10.2 Care of Film and Digital Storage Media

The film you use to take your astrophotos is the very material on which your original images are acquired, so if your film is damaged, so are your photos. Undeveloped film can be damaged by heat, moisture, or X-rays either before or after exposure. In addition, too long a delay (more than a few months) between the time you shoot your pictures and develop your film can allow the latent image to degrade, resulting in color shift and/or an effect similar to underexposure.

Developed film is relatively delicate and sensitive to humidity, so it should be protected in nonabrasive sleeves, kept in a dry place no warmer than room temperature, and handled only when you want to print or scan it.

Film and the emulsion on which your image is recorded is relatively soft, so due care is in order if or when you clean it. You would be amazed at how easily emulsion can be removed from film. Generally, the emulsion should never be rubbed when wet, though if you are careful, you can use a soft and lintless cloth or paper moistened with film cleaner to remove finger prints and the like. If you do accidentally fingerprint your film, it should be cleaned with film cleaner as quickly as possible. Oil and other chemicals in a finger print can visibly etch photographic emulsion in mere minutes, and some fingerprints can become more or less permanent if left on the film more than a day or so.

Due to etching from body oil, you should never rub oil from your nose on film (as some photo magazine articles recommend) to try and mask scratches on the film when making prints. If you have a badly scratched negative and don't have digital or other means of retouching images made from it, you should use more inert material such as the products made specifically for masking scratches on film, rather than sliming your film with oil from your nose. For the same reason, you do not want to breathe moist air on a negative when cleaning it. Some say that the mist in people's breath is like distilled water. This could not be farther from the truth. In reality, moisture in breath includes small amounts of mucous, microbes and other slimy stuff. If you can smell it, it's not pure water.

Magnetic media such as what you may use for digital images are sensitive to heat and magnetic fields, and some media are also sensitive to humidity and mechanical shock. Most types of magnetic media (such as floppy disks or Zip disks) may only reliably retain data for up to about 10 years, while magneto optical media may retain data for more than twice as long.

Magnetic hazards can lurk even at home. Of significant concern is the motor in a vacuum cleaner. You may think your magnetic media is all safe and sound when it is stored in a cabinet, but if the cabinet is not ferrous metal and your material is close enough to floor level that the motor of your vacuum can get within few dozen centimeters, your media could be at some risk. I personally heard about a case when the magnetic field from a vacuum cleaner damaged audio tapes. I have not yet heard of it actually happening to digital media, but if audio tapes can be damaged in this way, it follows that the same thing is at least possible with digital media. Other precautions include just using common sense about where you store magnetic media. For instance, you would not want to set disks under a telephone ringer or on top of a power supply or magnet.

A CD is not affected by magnetic fields and is usually more tolerant of other environmental factors. Commercially recorded CD media with a true gold coating for the data is better than media having aluminum, partly because aluminum coating can lose data due to oxidation if the CD's plastic surface is cracked. Some CD-R media uses dye or a combination of metal and dye, so a gold color does not always mean that the writable surface is made of gold. A DVD will hold many times more data than a CD, but the archival quality of DVD media has not yet been reliably established.

10.3 Identifying and Correcting Common Problems

The experience of shooting and seeing your first astrophotos can be satisfying even if some of them are not as good as you would have liked. If your first pictures were not all that great, you may eventually want to go out and shoot better photos of the same subjects. One challenge related to this is knowing what can be done to improve on your first photos. Factors which contribute to some image defects are obvious, while others may not be so easy to determine.

Fortunately, you can use your first pictures to evaluate factors such as your techniques, equipment, film, and exposure accuracy. Transparencies (slides) are the best for such test photos because you can tell right away if a slide is properly exposed. It can be difficult to determine the accuracy of your exposure by looking at a negative, and a print is not much better for this purpose because a photo lab can (and usually does) compensate for some degree of exposure error by varying the exposure of the print.

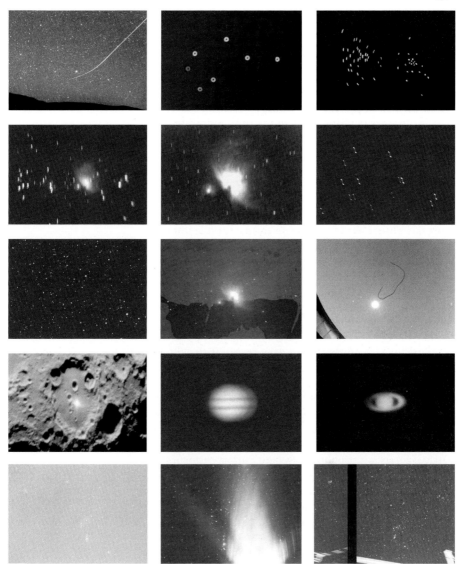

Figure 10.1. Some common astrophotography errors are shown here. Top row, left to right: airplane lights in picture; telescope out of focus; field rotation, typically from poor polar alignment. Second row: completely uncorrected periodic drive error; trailing due to inadequate guiding; accidental switching between two guide stars, flexure in guide scope mounting, or mirror slop in an internally focusing Cassegrain guide scope. Third row: underexposure or picture of wrong part of sky; ice or condensation on film emulsion during cold camera exposure; shadow of dust which was on film during an exposure. Fourth row: hot spot caused by reflection between an auxiliary optic and the secondary mirror of a Cassegrain telescope (or a rare case of unwanted reflections between optical surfaces in the auxiliary optic itself); poor collimation, centering, or tracking; subject or camera decentered during afocal exposure. Bottom row: overexposure or film too warm while being hypered; light leak in camera or localized film fogging during processing; film cut in wrong place.

Tracking or guiding errors are relatively obvious because of their characteristic streaked star images. Most flexure errors (usually from a guide scope mounting) result in streaks which run predominantly in one direction. These streaks can have a constant width or look more unusual. Tracking errors caused by mirror slop, focuser slop, or accidentally switching between two guide stars is more likely to result in a double star image instead of a simple streaked one.

Poor guiding can result in streaked images having an amazing variety of attributes. Such errors can cause a star image to look like a straight line, a thin streak which extends from one side of an otherwise round star image, an irregular squiggle, a plus sign, a round or irregular blur, a bow tie shape, or even a double image.

Field rotation due to poor polar alignment can cause streaking that is concentric with your guide star. You can tell if this is the problem by seeing if the star images are streaked more on the side of your photo that is farthest from your guide star. If your images have field rotation, the only good solution is to be sure your telescope mount is accurately polar aligned on your next outing.

Focus errors are relatively easy to see if the photo is properly guided, particularly if the telescope is a reflector having a central obstruction. A star image from such a telescope has a donut shape if it is sufficiently far out of focus, though the donut may only look symmetrical near the center of the picture. Lesser focus errors are more difficult to quantify, but a good way to verify if focus is a problem is to take a short exposure of the moon, a planet, or a bright star with the same setup you used for your blurred deep sky photo. This will allow you to eliminate focus error as a possible cause.

Some types of guiding errors can make a photo look fuzzy without producing the customary streaked star images. Other factors which can contribute to fuzzy images are poor seeing, internal tube currents, or high frequency vibration of the telescope. The latter problem is most common during high wind conditions.

Underexposure can be a show stopper for deep sky photography because an underexposed picture of a nebula will look more like a picture of nothing. The correct exposure for a deep sky photo can vary according to atmospheric conditions and how many degrees above the horizon your telescope is pointed. At many sites, you may have to double your exposure if the elevation angle of your subject is 8 degrees or less and quadruple the exposure if the subject is less than 5 degrees up. Deep sky photography can become impractical at such a low elevation angle because the ever present sky glow tends to be brighter near the horizon. Thin clouds or fog can also affect your picture; even if they are not obvious visually. Thin clouds or fog can cause your pictures to be underexposed or have an apparent glow around brighter stars.

Optical quality is another consideration. Some telescopes just won't produce a decent image no matter what; the worst case I ever saw was a 10 cm f/10 Cassegrain which had such bad spherical aberration that it would not resolve detail any finer than about 30 arc seconds. This is over 20 times worse than the theoretical limit and translates to a spatial resolution of $\frac{1}{7}$ mm on the film – over five times worse than even a mediocre camera lens! Such a telescope will not even resolve the rings of Saturn.

Other telescopes are well made, but they won't provide sharp images unless they are properly collimated. Fast Newtonian telescopes may have visible coma toward the edge of the image which can only be eliminated by using a coma corrector or stopping down the aperture.

Some telescopes have field curvature. This produces an image in which the center of the picture is in focus but the edge is not or vice versa. Here, you can try to average the focus between the center and the edge of the picture or get a field flattener. The usefulness of a field flattener depends on its design. Some may work well, but a marginally designed or incompatible unit may introduce its own visible aberrations.

Some of the most unusual looking errors relate to eyepiece projection and afocal photography. These errors can include a hot spot in the center of a picture (from unwanted reflections); dark spots on the picture (from dirt on the eyepiece optics or camera lens); fuzziness or color fringing on just one side of a planetary image (from decentering or other factors); or just an inability to get a sharp image. Using a different eyepiece or being sure the one you use is clean will usually clear up a lot of eyepiece projection and afocal photography errors.

Some errors may have nothing to do with the astrophotographer. Such errors include improper film hypering by another party; the rare case of just getting a bad roll of film; or when your film is fogged, scratched, incorrectly cut, or improperly developed by a commercial processor. Fogged or improperly developed film can be spotted in an instant because any visible defects will be evident on the part of the film which is outside of the picture area. In the case of 35 mm film, it is not unusual for fogging to cause occasional images of sprocket holes to appear on your film. If the film is improperly developed, it will all appear to have the wrong density or color. With 35 mm film, some forms of improper development can also be detected as diffuse lines which extend into the picture area from each sprocket hole.

If your negative film is completely blank or your slide film is completely dark, the absence of frame numbers along the edge will indicate that it was improperly processed beyond a reasonable doubt. If the frame numbers are present, then the fault is not usually with the photo lab.

Some types of fogging result from a light leak in the camera. Leaks which involve the camera back usually fog the film all the way to at least one edge. Other types of light leaks are the result of mediocre design of a camera's internal baffles. A case in point is when your film can be fogged if you simply change lenses during daylight. This type of fogging occurs only within the picture area and relatively close to one side because the stray light usually reaches the film from below the focal plane shutter curtains.

If the cause of various errors should elude you and your pictures keep having the same defects, a last resort is to try making arbitrary minor changes to your setup just to see if it makes a difference. As you get better at astrophotography and work the bugs out of your equipment and techniques, you will eventually be able to count on your pictures coming out like you want most of the time.

When you know why certain equipment and techniques facilitate their corresponding results, it will be easier to improvise for each unique situation. At this point, astrophotography techniques may become almost second nature, allowing you to get good photographs of various subjects and events with relative ease; all while focusing less on procedure and more on just having a good time.

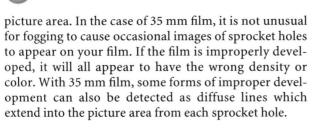

10.4 Image Processing Through Copying

Copying photos with conventional film typically increases contrast. This can be useful for emphasizing features on the moon and planets. Slow slide film usually

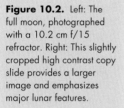

Figure 10.2. Left: The full moon, photographed with a 10.2 cm f/15 refractor. Right: This slightly cropped high contrast copy slide provides a larger image and emphasizes major lunar features.

provides the greatest useful increase of contrast because its base is typically darker than that of a faster slide film.

When certain films (particularly slide films) are used for copying, the result can also include increased color saturation which can provide very impressive nebula images. Copying can also bring out detail in bright or dark parts of the image, but this usually results in a loss of detail in parts of the image that have substantially different densities.

The down side to increasing contrast is that you tend to lose detail in parts of the subject which are not optimally exposed. For example, your copy of the main body if an emission nebula may come out as a deep red, but dimmer parts of the nebula may not be visible at all in the image. In addition, the contrast gain may cause star images to appear excessively enlarged or lose their subtle colors. In the case of a planetary image, increasing the contrast can excessively accentuate limb darkening.

Contrast in a copy can be reduced by introducing the appropriate type of flare. One way to get this flare is to use a thick piece of glass between your camera and the film or print you want to copy, with the glass angled at about 45 degrees from the optical axis. The entire copy setup is then oriented so an extended, diffuse light source can be reflected from the angled glass into the camera lens. The glass should be thick because this will help reduce sharp ghost images in the copy image.

Copying can also be employed to enlarge images. This can be particularly useful if you want to emphasize a given part of your picture in a slide presentation. An enlarged copy may not result in the best quality, but it at least provides a way to get a larger image of your subject if you lack long focal length optics. Some obvious flaws in a copy slide may include star images which are more bloated than those formed in an original photo through a longer focal length lens.

10.5 Image Processing in the Darkroom

Figure 10.3. Left: Original image of the area around M31, photographed with a 105 mm lens. The exposure is 30 minutes at f/2.5 on ISO 200 slide film. Right: Enlarged copy slide of M31, providing a 4× larger image.

Until recently, a darkroom was the primary arena in which amateur astrophotos were enhanced. Using an enlarger to print your own pictures allows you to retain control of your original film, which can reduce the risk of it being scratched or fingerprinted. Another advantage is that you can usually obtain the exact type of print you want; however, if you are picky about your results, getting the print you want may require a considerable amount of time and materials. If you make color prints, you can easily spend more on materials than the cost of having a commercial print made, but if you are a skilled printer, the final print you make has the potential of being just what you want.

10.5.1 Black and White Printing

Printing techniques such as dodging (shadowing a part of the print as it is exposed) and burning (locally exposing only part of the image more than the rest) facilitate local control over any desired part of the image. These techniques are simplest to implement on black and white images, but are also useful in color printing under the right conditions. Another benefit of black and white printing is that you can use a safelight which is bright enough to allow you to observe the print as it is being developed.

Variable contrast black and white paper such as Kodak Polycontrast™ and other media further enhance these techniques by effectively allowing one contrast setting to be used for a dodged or burned area and another for the rest of the image. This can be done by using filtration for a different contrast during the dodging or burning operation. The proper filtration can be provided by a special set of filters, or, if you have a color enlarger, setting equivalent values on its color head. Low contrast numbers produce low contrast prints and high numbers produce high contrast ones. A "normal" contrast setting is around 2.0.

Using a color head for Polycontrast settings can be handy because it will allow you to change the settings during your exposure, allowing you to use one contrast setting on an area that is burned in, then change it during the time the rest of the print is being exposed. Color head settings equivalent to Polycontrast filters are not always the same with every brand of color head due to differing calibration of the color head or the values of their dichroic or other filters. The table below is a good average between the values of brands like Beseler and Omega. A contrast of 1.5 has the least filtration and requires the least exposure, while other settings require more. Table 10.1 shows the exposure values (in terms of f/ratio or time) relative to the contrast 1.5 setting.

Table 10.1. Color head filter settings for black and white Polycontrast (TM) paper

Contrast value	Filter settings for Yellow	Magenta	Cyan	Relative f/ratio	Relative exposure time
00	1.50	–	–	1.3 f/stops	1.7
0	1.00	–	–	1.1	1.2
0.5	0.80	–	–	1.0	2.0
1.0	0.20	0.05	–	0.3	1.3
1.5	0.10	0.15	–	0	1.0 (reference value)
2.0	0.03	0.25	–	0.3	1.3
2.5	–	0.40	–	0.6	1.6
3.0	–	0.70	0.03	1.0	2.0
3.5	–	1.10	0.05	1.7	3.2
4.0	–	1.70	0.10	2.7	6.4
4.5	–	1.10*	0.40	3.0*	8.0**
5.0	–	0.70*	0.80	3.4*	11.0**

* Some types of photographic paper require far more magenta filtration at contrasts above 4.0. The most extreme is 3.20 magenta for contrast 4.5 and 5.00 magenta for contrast 5.0. This is more filtration than most color heads provide, so supplemental filters may be required. At these filter settings, the required exposure for contrast 4.5 can be up to 16 times greater than that required for contrast 1.5. Contrast 5 may require up to 40 times more exposure than 1.5.

** Estimated maximum exposure change for contrasts 4 and 5 at shown filtration. Actual required exposure compensation may vary substantially (several stops) at these contrasts depending on the subject matter.

10.5.2 Masking, Sandwiching, and Other Techniques

More sophisticated techniques such as sharp or unsharp masking provide even more flexibility and repeatability. Depending on the subject, the mask can either be derived photographically from the original image or be artificially created or modified with dodging or burning techniques. The mask can be used for anything from a simple dodging mask (eliminating the need for manual dodging or burning), to a contrast or sharpness mask (where an unsharp image (sometimes one having low contrast and inverted density) is sandwiched with the original in order to locally modify density or enhance the contrast or range of detail in small features).

Sandwiching of an original image and its mask do not necessarily have to occur in the negative carrier; the mask can instead be a larger piece of film which is contacted with the print media during the exposure. I prefer this approach because it can facilitate more accurate registration of the images. Film suitable for this purpose can usually be obtained from some graphic arts suppliers. Some types of this film can also be used to make positive or negative transparencies instead of prints.

Other techniques include sharp positive or negative masking, where a sharp positive or negative mask of the image is used to raise or lower contrast. Sharp or unsharp masks can also facilitate composite exposures where different pictures of subjects are combined.

Composite (or integrated) exposures can also involve sequentially exposing multiple negatives (or transparencies) of the same subject onto a single print in order to suppress the apparent graininess of the final image. This works because the subject matter is common to all of the negatives, but the film's randomized grain structure is different in each image. The method is also useful for incremental masked printing of subjects having a high dynamic range.

Pin registration is useful for composite images because it will allow you to interchange the print paper with a focusing surface which is marked to show features in the subject you are using for position references. A vacuum easel is helpful when you use print masks because it can keep the paper and mask flat. You can build an effective vacuum easel by building a flat wood box having a pegboard top, then sticking the

Figure 10.4. Top: Elaborate images can be made through composite printing in the darkroom or equivalent forms of digital processing. This composite portrait was made entirely in the darkroom by printing three separate pictures onto a single piece of photo paper. Simple cardboard cutouts were all that were needed to facilitate the extensive localized dodging and burning that was required. Middle group: These three images were combined to make the top circular composite portrait. Bottom: This composite print was made from five separate pictures and required several pin registered masks. The fisheye picture shown above provided the foreground and sunrise; the above sun ray photo includes the clouds on top; the mountain is from a third picture; the first two horse riders are from a fourth picture, and the remaining horse riders are from a fifth. The ranch logo was added by the brochure's printer. I made these composite pictures in my darkroom less then a year after graduating from high school in 1976. If I could produce these images in a darkroom when I was a mere teenager, one can certainly take on work of similar complexity with today's digital image processing tools.

hose from your vacuum cleaner into a hole in the side. If you have an excessively strong vacuum, you may want to add some small holes in the side of your easel or make it oversize so several peg board holes will be unobstructed when the photo paper is placed on the surface. This will lessen the easel's vacuum effect to a manageable level.

Other easel enhancements include a raised border to accurately register the print. This makes it easier to use the easel for borderless prints without using a vacuum. A suitable raised border should be dark and rough so it will not obliquely reflect light from the enlarger onto your print. You can also add shallow grooves which will allow you to get your fingernail or a thin ruler under the paper in order to remove it from the easel more easily. Painting the printing surface black can help prevent light that gets through the print media from reflecting back toward the emulsion. To focus your enlarger, just place a sheet of white paper on the easel's printing surface. This type of home made vacuum easel can be very effective. After I built mine, I never wanted to use a commercial easel again!

Sandwiching two or more negatives or transparencies is another technique which can reduce the apparent graininess of the final image. An additional effect of this method is increased contrast. Unlike sequential exposures, sandwiching results in an effective multiplication of the densities of two or more original or copied images. The effect is easy to envision when you consider the amount of light that makes it through each sandwiched image. For instance, say two identical film images are sandwiched. For simplicity, we'll say that the brightest area of each transmits 60 percent of the light and the dimmest part transmits 2 percent. Therefore, the density range of each film is 30:1; the equivalent of just under 5 f/stops. Accordingly, 60 percent of the light gets through the brightest part of the first film, and 2 percent gets through the darkest part.

The second original film has the same characteristics, but less light reaches it through the first film. Further, more light reaches the lightest part than reaches the darkest part. Therefore, 36 percent (60 percent of 60 percent) of the light will reach the print paper through the brightest part of the sandwiched films, but only 4/100 of a percent (2 percent of 2 percent) will get through the darkest part. Squaring the dynamic range ratio of one film gives you the overall ratio for two identical sandwiched films. In this example, sandwiching multiplied the dynamic range of the original image from 30:1 to 900:1! To determine the new dynamic range in f/stops, just add f/stop units: 5 stops + 5 stops = 10 stops.

Sandwiching is useful for many types of film, including identical thin negatives, overexposed slides, or sharp masks made from either. The effects are most dramatic on films having a low base density and some resistance to "blocking" of overexposed highlights. The dynamic range of a sandwiched set of films can easily exceed what most photographic paper can encompass, so the technique is most effective when you want to get a print that emphasizes subject matter within a particular density range.

10.5.3 Developing Black and White Prints

The developer you use can control many attributes of a black and white print, but most special developers have more to do with the warmth of a print (how much toward yellow or brown it is) rather than attributes such as contrast. The overall developing time can be used to control contrast and density to a degree; however, this is not recommended as a substitute for using the proper exposure value. The best image will usually result from a properly exposed print and the recommended developing time. If the development time is too short, a cloudy looking picture can result.

Some things to guard against include contamination of your chemistry. If you get so much as one drop of stop bath or fixer into your developer, it can cause visible problems. It is important to adequately drain each print before moving it to the next chemical or to rinse it between each chemical bath. In some cases (depending on the process), you may be able to use a rinse in lieu of a stop bath. It is usually a good idea not to save used developer or stop bath, and it can be helpful to use two fixer baths. Try to avoid the temptation to turn the light on as soon as your print hits the fixer. Turning the light on too soon can cause problems that may result in premature yellowing of the print.

Adequate washing after fixing is very important, but there is a difference between a good wash and a long wash. Your print washer should provide for good circulation of the water and you should move the prints around often, being careful not to scratch them. If you are using resin coated paper, the wash time should not be too long, since too long a wash could cause the paper to separate. Ideally, all the chemicals and washes should be relatively close to the same temperature; say within 5 degrees Celsius or so for black and white and within less than half that range for color.

How to dry a print is dependent on the type of paper you use. Conventional paper can be dried in a blotter book, but the best results are obtained by using a print dryer. Most types of paper should be placed on the dryer with their emulsion away from the metal surface. The lone exception is when you want to get glossy prints with "F" (glossy) paper which is not resin coated. In this case, the emulsion should be placed against a polished metal surface such as a ferrotyping plate or the surface of a dryer specifically designed for the application, then gone over with a squeegee or roller to ensure good contact. Typical resin coated paper can be air dried or placed in a blotter book, but it should never be ferrotyped.

10.5.4 Manipulation of Black and White Prints During Development

Manipulation of a black and white print is possible during development. One method I learned from a news-paper photographer involves lifting at least part of the print out of the developer and lightly rubbing the part(s) you want to darken with your fingers. The technique is effective because developer works faster at the higher temperature of your fingers. It is important to keep dipping your fingers into the developer so the print will stay wet and to feather the darkening effect by occasion-ally rubbing the area immediately surrounding the part you want to affect. In addition, the entire print should be placed back into the developer every few seconds so you don't get streaks from uneven distribution of developer on the rest of the print during development.

You can get more extreme results by rinsing the par-tially developed print in water and then rubbing devel-oper onto the area you want to darken. Your hands must be clean when you perform this procedure in order to prevent contamination of the developer or staining of the print. Obviously, this technique should only be used if you are certain that getting developer on your skin won't cause an allergic reaction or other problem. I have not tried the technique with water proof gloves, but it may work if the glove material is thin enough.

10.5.5 Color Printing

Color printing is a lot easier than it used to be. When I started making color prints more than two decades ago, some processes required two developer steps and exposing the print to a photo flood lamp part way through the process. Further, the chemistry temperature had to be within about 1 degree Celsius. Today, you don't have to expose the print to a photo flood lamp, only about half of the chemical steps are required, and the chemistry can be used throughout a wider range of temperatures.

Some of the image manipulation techniques mentioned above can be utilized in color printing; however, color printing imposes a number of limitations. For one thing, you can't use a safelight bright enough to permit evaluation or manipulation of the image as it is developed. I don't use any safelight for color printing because some can cause a slight magenta cast in highlight areas of color prints from negatives. Panalure paper (a black and white paper made specifically for making black and white prints from color negatives) is also sensitive to most safelights.

Color paper is also very sensitive to white light such as that which may enter through an improper door seal or leak from your enlarger and reflect off the wall. With color negative paper, modest white light fogging typically causes a slight cyan cast in highlight areas of the print. More severe fogging may look gray or red. Fogging of positive paper usually causes dark areas on the print to be gray or take on a colored cast. Effects similar to the appearance of fogging can also result from improper development or fixing, staining by old or contaminated developer, or other chemistry problems.

A second consideration is that reciprocity failure causes the color balance of most color paper to shift according to exposure time. This can require appropriate filtration changes if you use different exposure times. It can even require local control of the filtration when you dodge or burn a print.

Yet another consideration is that the color response of photographic paper shifts as it is stored over time. With some paper, a noticeable color shift can occur in as little as one day when the paper is at room temperature.

Color shift over time is even more pronounced after a print has been exposed. In the case of some paper for printing color negatives, I can see a slight shift toward green even if the variation in time between exposure and development is only half an hour. This can rule out saving up batches of exposed prints to develop all at once unless you adjust your filtration to compensate for the variation in time between exposure and development.

In most home darkrooms, color prints are developed in drums rather than open trays. This facilitates relatively consistent results because a small amount of fresh chemistry is all that is required for each print. Prints processed in drums can be susceptible to chemical stains (particularly at the edge) but I have found that this problem can be virtually eliminated by manually rolling the drum instead of using a motorized agitator. Manual rolling allows me to occasionally rock the drum back and forth at positions where the paper edge is bathed in the chemistry. Other things that help are to use at least 70 ml (2.5 ounces) of chemistry per 20 × 25 cm print and to add a brief water rinse between each chemical step. I like to use at least 220 ml (8 ounces) of water per 20 × 25 cm print (or equivalent) in each rinse step.

Drums are available in many sizes, and most larger ones are designed to accept one large print or several 20 × 25 cm prints. Some will even accept multiple 13 × 18 cm and 28 × 36 cm prints. Test prints can be developed in a small drum to save chemistry, but there can be some variation in print density according to the brand and size of each particular drum. If the drum, wash, and chemical temperatures are consistently maintained, the variation can usually be predictably compensated for when exposing prints which will be developed in different drums.

Drum processing works best when the paper is not cut. This minimizes the risk of small pieces of paper shifting and covering each other during processing. You can make several small prints and test prints on a single piece of standard size paper by making cardboard masks which cover all of the paper except the part you want to expose. These can consist of a single mask assembly having several hinged panels which are selectively flipped up, or, a set of separate fixed masks which are used in various combinations. I prefer the fixed masks because they reduce the risk of fogging parts of the paper I don't want to expose at a given time. The masks should be dark in order to minimize how much light gets reflected toward the enlarger, darkroom walls,

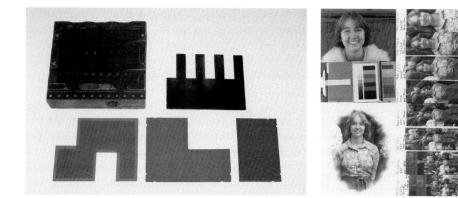

etc. Cardboard is more than adequate for such masks if you take care of them – I'm still using the same cardboard masks I made more than two decades ago!

10.5.6 Hybrid Techniques

Some darkroom techniques may involve multiple generations of procedures, where the print you expose from your original film is not the final result. This can be particularly true if you need to gain a lot of contrast or if you want to implement dodging or burning techniques which are too complex to manage during a single print exposure. A total solar eclipse is a very demanding subject because the solar corona is over 1,000 times brighter near the center than it is at its outer visible extremes. Accordingly, a straight print (particularly from a slide) will only reveal detail within a narrow annular zone surrounding the moon; features inside the zone are usually fried and features outside the zone look too dark.

Visually, the entire corona typically looks the same pearly white color, but many color films may add false colors in dimmer areas. Therefore, there is no real loss in utilizing black and white media for all aspects of an eclipse image other than the pink prominences and possibly the grayish blue background sky. By printing a slide of the corona on black and white paper and radially dodging the print to gradually increase its exposure toward the edge, the range of detail from any given slide can be radically increased. The resulting print is a negative, but you can get a positive print by copying the print with slide film, then printing the resulting slide on the same type of paper. Three or four such

Figure 10.5. Left: Cardboard masks which can be accurately registered with your easel facilitate exposing several small prints on a single sheet of paper which can be processed in a standard drum. The upper left object is a vacuum easel made from pegboard and a few scraps of wood. Right: This 20 × 25 cm print was made by using the vacuum easel and various combinations of the cardboard masks shown at left. It includes two 10 × 12.5 cm prints and eight 3.1 × 9 cm test exposures; all printed on a single piece of color photo paper.

dodged images from the same number of sufficiently different original exposures of an eclipse can capture virtually the entire range of observable brightness.

It is also possible to get a quick (though possibly mediocre) black and white print from a slide by printing it backwards onto black and white paper, developing the print, then placing the print face down (under glass for good contact) onto another piece of print paper and exposing it with light from your enlarger. The first print is a negative, but when it is contact printed, the final result is a positive print. The success of this technique is dependent on the use of paper that does not have dark letters (such as in a logo) on the back, since a negative image of any such letters will be reproduced on the final print. A thin paper backing is also better than a thick one for the negative print.

10.6 Analog Processing of Video and Still Images

Analog processing of images with a video color corrector or other electronic equipment is more applicable to viewing an image than getting a final print, but it can occasionally bring out details that can't be seen on reflective prints or brought out with most other types of image processing. This may be due in part to the fact that a television or monitor is a light emitting device which is not dependent on ambient lighting conditions. All of the resulting image detail may not show up on a printout, but it can provide a nice video tape!

An example is one of my original black and white negatives of the November 1994 total solar eclipse. It captured 3 degrees of a coronal streamer; so much length that it went to the edge of a 350 mm lens' coverage! Before trying to print a picture, I sometimes preview it with my video camera by using a close up lens and the front of a slide copying attachment. This allows me to capture a raw video image of the negative which I then convert to a positive using a Sony XV-C900 video color corrector. This unit allows me to fine tune color balance, color saturation, gamma, and have some apparent control over contrast by partially fading to black or white. Among other things, this allows me to quickly see what detail is available on a negative. It also allows me to implement special effects such as adding a slight blue color to the sky in a black and white solar eclipse image.

By using the color corrector unit to process a video image of my eclipse negative, the coronal streamer was obvious all the way to the edge of the negative; however, when I printed the picture in a darkroom, I could not get the outer extent of the streamer to print, even when using dodging or burning techniques. Digital image processing from a scan of the negative did not bring out the detail either, but this could have been due in part to the limitations of the 24 bit film scanner I was using.

A frame grabbed from the video processor output retained most of the corona, but suffered from the low resolution characteristics typical of video. Fortunately, characteristics other than resolution can be just as important for some subjects, and in cases where images will be combined digitally, even resolution limitations can be overcome if you shoot a close up video in which you slowly pan across a wide field subject, then assemble a mosaic which encompasses several adjacent frames.

If you lack a film scanner, a video camera can be used along with your computer and a frame grabber to "scan" some of your photos. The full frame results are not as sharp as those provided by a good film scanner, but they are certainly adequate for a snapshot such as you might display on a modest web page. I used a camcorder and close-up lens to "scan" some last minute photos for this book. For example, the four images of the moon occulting Jupiter in Figure 9.3 were "scanned" with a camcorder and a frame grabber.

A suitable video processor can increase color saturation of the video from which you grab frames. This provides good color saturation in the original grabbed frame, which in turn can tend to reduce the harshness of any visible pixilated artifacts in your final image.

If your video camera does not focus close enough to get a full frame shot of your film, you can try using some photographic close up lenses in front of your video camera. An even better solution is to use a reversed achromatic objective lens from a 30–50 mm aperture binocular or finder scope as the close up lens. Such a lens can provide a sharp close up image as well as allowing you to use the zoom lens of your video camera to effectively crop the image.

10.7 Digital Image Processing

PC based digital image processing has become the most popular form of image processing among many amateur astronomers and photographers. Image processing is a subject worthy of its own book, so only a few basics about major capabilities of popular image processing software will be covered here.

Unlike darkroom based work, digital image processing permits you to easily replicate your work in various stages of completion so you don't have to start from scratch every time you make a mistake or decide to change a step. Image processing programs such as Adobe Photoshop allow you to easily manipulate image attributes such as density, contrast, color balance, and color saturation. It also allows you to selectively influence these factors within limited areas or density ranges of the image.

Adobe Photoshop and some other image processing programs also make it very easy to eliminate flaws such as dust or scratches by either influencing the sharpness around small features or manually "cloning" part of an image over any defects. Many image processing programs also have features which produce effects similar to darkroom based unsharp masking and other techniques; however, not all digital image processing routines are strictly analogous to the darkroom processes having the same customary name.

Many problems common to various astrophotos can be overcome with digital image processing. An example is the false color which plagues many planetary photos. Even if you use a reflecting telescope, decentering of the planet or imperfections in the eyepiece you use for projection or afocal photography can add false color to your image. This problem is most often detectable as color fringing which is red on one side of the planet and blue on the other. Since lateral chromatic aberration is usually the culprit, a color scan of the original image can be corrected by independently moving the red and blue primary components of the image (each called a "channel" in Adobe Photoshop) until they are registered to the green component. This not only eliminates the color fringing, but it can also make the entire image look sharper because its component colors are now properly registered! This process takes only a few minutes on a computer, but getting even close to the same results in a darkroom can involve an arduous series of procedures which add up to many hours.

Images that do not have excessive grain can be sharpened with the "unsharp mask" filter. This is particularly useful for increasing the zip of photos having a lot of star images. Too much application of the filter can cause a dark ring around each star image. Fortunately, the harshness of this type of ring can be reduced through selective application of the "curves" function, a tool which permits selective adjustment of an image within a selected range of brightness. The curves can also be used to minimize the harshness of film grain or increase the brightness of a nebula image without affecting brighter or darker features. Another use of the curves function is to suppress flaws such those caused by flare, light leaks, or undesirable optical reflections.

Digital image processing can also be very useful for getting good black and white images from color originals or adding false color to black and white images. Color negative film in particular can have a very low contrast in terms of absolute density. Therefore, some features which are obvious in a color image of a planet or deep sky object may become washed out when the unmodified image is converted to black and white.

I repeatedly encountered this problem while converting my color images to black and white for this book; some features which were obvious even on a machine print from the photo lab became very washed out or even disappeared when I converted a scanned file of the same image to monochrome. Accordingly, a little image processing was required in order to recover features which were obvious in the color originals.

In general, this processing consisted of using the "Gaussian blur" filter to blur a typical planet image up to 2 percent of its diameter in two of its primary colors, increase the contrast in at least those two colors, then convert the image to monochrome. The result shows about the same detail as the color original. Increasing the contrast of a black and white image after conversion to monochrome was less satisfactory, since this also increased the apparent harshness of the film grain.

Original black and white images shot with high contrast or spectrally selective black and white film can easily provide final black and white photographic prints and digital scans which are better than the black and white conversions of my color images in this book. In addition, combining a high contrast monochrome image and a color image can provide a truly outstanding composite color image.

I used my existing color photos throughout most of this book rather than going out and shooting new black and white pictures because most amateur astronomers (including myself) like to shoot color astrophotos and I wanted the published results to be representative of what one can expect from popular color films.

Deep sky images such as the one of the North American nebula in Chapter 2 also required some processing in order to get red nebulosity which is obvious in the color original to be equally visible in the published black and white image. This processing consisted mostly of adjusting the brightness and contrast of the red channel prior to converting the image to monochrome, but only to a degree that allowed the final monochrome image to reveal about the same degree of nebulosity as a standard color print from the original negative.

By using only the red channel and further increasing the contrast, it would have been possible to make the nebulosity look a lot brighter, but I did not want to do that for the images in this book because such a result would not be representative of what one could expect with the exposure time I used.

Digital image processing can make it relatively easy to combine multiple pictures into a single image which combines the best attributes of each. This is useful when combining varied exposures of a subject having a wide dynamic range. Such subjects include a total solar eclipse or a field of view which contains both the moon and a deep sky object.

Other applications include those in which multiple images of a subject are overlapped in order to provide a final picture with less apparent grain or other flaws than any one of the individual images. This can work to some degree because the features on the subject are common to all the images, while grain and attributes such as atmospheric seeing effects are all a little different in each picture. The ring nebula images in Fig. 4.3 show a limited application of this technique to video images.

If you are combining multiple images of a planet, the entire series of pictures should be taken when it is reasonably close to the same rotational orientation. The rapid rotation and dynamic atmospheric conditions of

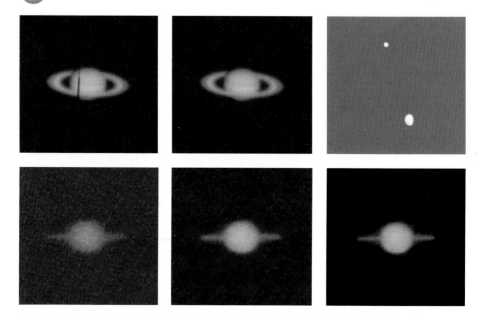

Figure 10.6. Top left: Remember the annoying reticle in the Saturn image back in chapter 3? Top center: Not a problem if you appropriately use of the clone tool in a program like Adobe Photoshop! Alternately sampling and cloning the image from each side of the linear reticle image allowed it to be removed in a matter of minutes. Top right: Digital image processing can be used to simulate events which were not even photographed. Here, two modified images of Venus were used to simulate the unusually close (well under an arc minute!) daytime conjunction of Mars (top) and Mercury (bottom) which occurred on 4 September 1985. This simulated image is based on a drawing I made of the event at 21:00 GMT. The image is reproduced here at about the same angular scale as the Saturn photos. Lower left: A small, underexposed slide image is one of the few photos I have of Saturn near the time of any of its ring plane crossings. This raw scan is pixilated due to the small submillimeter size of the original image. Making an extreme enlargement in the darkroom and scanning it with a flatbed scanner would have been vastly superior to this scan of the original slide. Lower center: A "Gaussian blur" filter was used to significantly blur the planet in two of its primary colors, then to blur it only slightly in the remaining color. After this, the contrast was increased and the image was converted to monochrome. This image looks about as good as the original slide, minus a few large grain clumps which were subdued by minor dodging and burning. Cloning would have eliminated the clumps entirely, but the result would not have been as true to the original photo. Lower right: Using extremes of the above process to make a high contrast mask layer and performing some significant retouching provides an image that looks radically superior to the original. The result actually goes a bit far in that it makes the rings look a little closer to edge on than they really were in the original photo. I did not go this far when processing images for the rest of this book because the object of this book's other photos is to provide published black and white images which have about the same degree of visible detail (and flaws) as the original color photos. The purpose of this example is to show that you can use image processing to salvage a bad original photo. If you start out with a good original photo, the results can truly be spectacular!

Jupiter may require you to shoot your entire series of pictures in only a couple of minutes. Photography of a planet having a slower rotation does not have to be as rushed, and a planet such as Mars can usually be photographed when it is at the same rotational position on consecutive days. Limb darkening on some planets will become more obvious as you increase the contrast of a single or composite image, but radial dodging can keep the limb from becoming excessively dark.

Some image processing programs make it easy to change the size and proportions of an image. Such changes can be as subtle as stretching part or all of an image to compensate for the characteristics of a given lens, to complete geometric conversions which affect the distribution of every element of an image. An example of the latter is the "polar coordinates" filter in Adobe Photoshop, which can convert a circular panoramic or all-sky image (such as that acquired with

Figure 10.7. Top left: This annular 360-degree "panorama" was captured in one shot with a vertically oriented Cassegrain axial strut wide angle reflector that I invented and built in the late 1970's. Top center: In mere seconds, the "polar coordinates" filter in Adobe Photoshop converted the annular polar image into a square image having rectangular coordinates. Top right: Cropping off the dark top of the square image (which corresponded to the dark spot at the center of the original annular image) and scaling the image in one dimension to normalize proportions provides this sweeping 360 degree rectangular panorama. The vertical angle of view is about 110 degrees. Lower left: This radially dodged image of the 3 November 1994 total solar eclipse is from a 1 second exposure at f/11 on Kodachrome 64 professional film, through a Vernonscope 9.4 cm f/7 refractor and a Barlow lens which was working at about 1.6x. The original slide was printed on black and white paper and radial dodging was used to widen the range of visible detail in the corona. The resulting negative print was converted back to a positive image with Adobe Photoshop. Lower right: Here is a new way to look at the corona. In mere seconds, the "polar coordinates" filter in Adobe Photoshop "unwrapped" the corona into this panorama. In this image, the radial image scale of the corona (relative to the horizontal or circumferential scale at the lunar limb) is significantly exaggerated in order to better show structure in the fine polar streamers. This geometric conversion technique can be used to enhance analysis and image processing of the corona, then the processed image can be converted back to its original circular form.

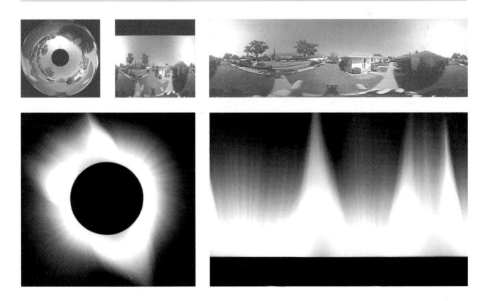

a vertically oriented wide angle reflector) to a straight rectangular panorama.

When converting an annular or circular image to a rectangular panoramic format with the polar coordinates filter, it is important that the area selected for conversion is completely square, and that the horizon circle is concentric with the exact center. Otherwise, the horizon will look like a wave in the panorama rather than being straight. For this application, the "horizon circle" would be the circle in the image which would describe a flat horizon or a level plane which intersects the vantage point in the picture. This would be different from the real horizon in areas having uneven terrain.

It is possible to make such drastic geometric conversions in the darkroom, though it is relatively time consuming. For example, it could easily take more than a day to properly convert an entire 360 degree panoramic image using a curved easel and related darkroom based process I invented and implemented back in 1976, while a computer based image processing program can do it in anywhere between a fraction of a second to a few minutes; depending on the computer, software, and file size.

10.8 Synergistic Combinations of Analog and Digital Image Processing

Digital image processing is useful for many things, but it still falls short of being everything anyone will ever need for every image. This is in large part due to the limitations of scanners. Some scanners allow you to adjust the actual scan exposure in order to provide the best quality in the density range you want. This allows one to acquire a single optimized scan; or, two or more scans (which would include one light scan and one dark scan) that have enough combined bit depth to encompass the entire density range of the original film. The scans can be combined when the images are processed.

One related problem is that many scanners do not offer a true exposure control. Instead, some perform all scans at the same exposure value, regardless of the "exposure" you set. In this case, the so-called exposure control may be nothing more than a canned image processing routine that changes the brightness attributes of the digital image file after the scan is completed. This is not as good as a real exposure control because it does not provide a way to increase the raw bit depth for the image data at either end of your film's density range, which is usually where you need it. Scans of underexposed or dark images (such as astrophotos) with this type of system typically exhibit a lot of artifacts from the scanner's sensor. This problem can be overcome to a degree by using a blank scan as a "flat field" reference.

Another flaw in some digital images is the appearance of visible contours which correspond to boundaries between different density values. Photographic film is a true analog system which can capture very subtle differences in color and density. Digital images are limited to a finite number of specific densities and colors. There may be millions of colors available, but the catch is that there may only be a few hundred levels of density available for any given primary color. This can result in contours in digital images which are most obvious in extended highlights or areas made up mostly of a single color. Such contours can become more obvious than usual if the image is blurred enough to eliminate the film's grain structure. In this case, some suppression of visible contours can result if you deliberately add some randomized noise (such as graininess) to the image.

Since digital image processing is subject to the above limitations, there is often much to be gained by utilizing additional image processing disciplines such as darkroom based work. I have found it very useful to make dodged prints or transparencies in the darkroom, scan them, then digitally process the images. By using darkroom techniques to lower the dynamic range or emphasize the attributes of interest, the final result can be substantially better than what I would get from digital image processing alone. Most of my work in this area relates to total solar eclipses rather than deep sky astrophotos, but even eclipse related work can show the effectiveness of hybrid techniques.

10.9 Special Effects

Image processing can be used to bring out real detail within a subject or to create special effects. Such effects can include combining a silhouetted scene with your favorite astrophoto or even creating images of observed events or fictional planets or nebulae. Special effects can also include 3D stereo images or color contours to show varying ranges of brightness.

10.9.1 Stereo Imaging

Stereo images of terrestrial subjects are typically captured by utilizing two or more separate vantage points. Most often, two cameras or synchronized shuttered lenses are used so both images can be taken at the same time. This allows factors such as lighting (and expression in the case of a portrait) to be the same in each image. Separation between the camera lenses will determine the apparent depth of this type of stereo image, and the appropriate separation is dependent on the depth of the subject and the angular field of view.

If your subject is only a meter or two away and you are using normal lenses, a separation about equal to that of your eyes is appropriate. If instead the subject is a mountain several kilometers away and you are using lenses so long that all foreground objects will be excluded, a camera separation of several meters may be in order.

Astronomical subjects such as meteor showers can also be recorded in stereo. In this case, the appropriate separation for cameras having wide angle lenses may range from a few dozen to a few thousand meters. You can also get a stereo image of the moon by coordinating with another astrophotographer to simultaneously shoot photos of it from widely separated (by thousands of kilometers) points on the earth.

You can also acquire stereo images of a wide variety of astronomical objects using a different technique. Rather than using separate simultaneous vantage points, you can let the motion or rotation of the earth or your astronomical subjects provide different vantage points for you. This can be accomplished by taking the pictures at different times. Such image pairs will provide a stereo image of the sun or a planet by capturing them at slightly different angles of rotation. Likewise, the technique can take advantage of the moon's libration to capture stereo images of it during different occurrences of its various phases. Other stereo effects are derived from the relative motion between two or more objects. Subjects for these techniques include a totally eclipsed moon in front of

Figure 10.8. Two images of certain astronomical objects which are taken at different times can provide a three dimensional stereo image. If you turn the book sideways, the left two enlargements from photos of the 30 December 1982 total lunar eclipse will appear three dimensional (the moon will look closer than the background stars) if you look "past" them until they appear to merge and then focus your eyes on them while still making them look merged. The center image and the right image will appear three dimensional if you look at them while slightly "cross eyed" so that the images appear to merge, then focus your eyes on them. If you don't find either method easy to do, don't try looking at the images so long that you strain your eyes. (A good percentage of people may not be able to get their eyes to accommodate either viewing method.) If this is the case with your eyes, you may be able to see the left two images in stereo if you use two identical but weak magnifiers (one in front of each eye) which allow you to view the images from a distance of between about 15 and 30 centimeters. The outer photos are enlargements from the same 5 minute exposure through a 105 mm f/2.5 lens on ISO 200 slide film. The center photo is a one minute exposure. Three dimensional imaging usually works better if both images in a stereo pair have the same exposure.

a star field, a total solar eclipse, or a comet. The apparent depth of stereo astronomical photos is typically determined by the field of view and the time between exposures.

In any stereo imaging, it is important to get a realistic 3D stereo effect. Too little stereo effect won't let you see any obvious depth in the subject. Too much effect and the apparent depth will be so extreme that your eyes can't accommodate the exaggerated depth of the subject, possibly resulting in eye strain. In general, the entire image should appear to have a depth of no more than about 5 percent of the viewing distance. In fact, less than half of this value works best for most astrophotos.

For a viewing distance of 30 cm and an interocular (interpupillary) distance of 65 mm, a 5 percent difference in apparent depth also corresponds to a 5 percent difference in the interocular distance. In this example, the 5 percent maximum relative variation (parallax difference) between foreground and background elements in each of the pictures is about 3.25 mm (5 percent of 65 mm), with the preferred reduced variation of 2.5 percent for astrophotos being 1.6 mm or slightly less.

A good way to get the effects you want is to take a series of pictures having different degrees of stereo effect, then pick the pair that looks best. If you later want to look at the image pair from a closer distance, the measured parallax difference in the images should be reduced to correspond to less than 5 percent of the new viewing distance.

Remember that the maximum parallax differences shown above are for the viewed print and not the original negative or slide. The image on your film will most likely be enlarged, so the measured difference between foreground and background objects must be less on the film than on the print. For example, a 35 mm negative must be enlarged 4 times to provide a 9 × 13 cm print. Therefore, a parallax difference of 1.6 mm on the print will require a difference of only 0.4 mm on the negative.

The absolute distance between viewed pictures depends on whether you will be using a viewing device or just skewing your eyes and looking directly at the pictures. Most stereo viewers (such as antique ones) are made for 75 mm square images and a separation of 76 mm between background elements of each image. Other viewers are made to work at a customary interocular distance of about 65 mm. Most of these viewers have magnifying lenses which shorten the viewing distance, and this should be taken into account when establishing your maximum separation distance. If your original images can be mounted in a standard slide mount, you can use two identical slide viewers to preview the stereo effect.

You can look directly at stereo images in two ways. One is to look straight "past" the pictures (i.e. straight ahead, so your eyes are parallel) and focus on the nearby prints. This method usually limits you to using prints on which the respective background features have a maximum separation equal to the distance between your eyes. Another method is to cross your eyes enough to make the prints appear to merge, then focus on the prints. The method you use will determine which print should be on which side in your stereo pair.

If you look past the pictures, the prints should be arranged so the foreground parts of the subject are closer together than the background. If you look cross-eyed at the pictures, the position of the prints should be reversed. If you want to use either viewing method, you can try using three pictures, with the outer two images being the same. This will allow the left and center images to be viewed with one method, and the center and right images to be viewed with another.

Rotational and lateral alignment of the images is also important because misaligned images may have an uneven stereo effect or maybe even give you a headache. Ideally, the viewed images should be oriented so that the relative offset positions between elements of the subject vary only along a line parallel to that between your eyes; that is, the offset should be to the right or left, not up or down. It is also a good idea for the boundary of each print to correspond to the same part of the subject background.

Looking at stereo images can cause eye strain and take some time to master. In order to avoid eye strain (and a possible headache) you should (at least to start out) only practice looking at an image set for a maximum of a few seconds on any given day. In addition, you should stop looking at the images (at least for a while) if any pain or discomfort results from trying to do so.

10.9.2 Other Effects

Digital image processing can be used for simple procedures like retouching dust spots, scratches, and larger flaws. Image processing software can also let you create images of observations that you did not get to photograph. Here, a drawing or a few notes about your observations of the event are very useful. Conjunctions are relatively easy to reproduce by "cloning" images of planets onto a black background layer and adjusting their size, position, brightness, and color to match the documentation of your observation. This may be the only required step if the occurrence was at night. For daytime events, you can select the background, then "paint" it to match the color of the daytime sky. Creating an image like this may be "cheating" in a sense, but it does at least look better than a pencil drawing of the same event.

10.10 Astrophotography and Astronomical Software

Many of the objects and events you photograph may be covered in various books and magazines, but there are times when such published material may not be enough. For instance, say you want to know what a certain astronomical event will look like from a given location or learn when an object rises or sets. Yet another relates to when you want to know about an upcoming occultation or conjunction before it is covered in a periodical.

There was a time when all you could do was look at a periodical or specialized book to find out when an astronomical event of interest would occur and where it could be observed from. The problem with this is that a periodical may only include information about events its editor thinks people will be interested in. Details about events such as eclipses can be gleaned from a book such as Jean Meeus' *Canon of Solar Eclipses*, but the reader must plot the path coordinates on a map in order to accurately determine the limits of totality.

In recent years, new tools have become available which put the amateur astronomer in the driver's seat in regard to the information he or she can get. One of these tools is the Internet. It provides a nifty way for amateur astronomers to share information directly with each other. Another tool is even more useful for some applications because it allows amateur astronomers to perform their own research without depending on specific published data. This tool is astronomical software.

10.10.1 Learning of Future Astronomical Events

Even the most basic astronomical software usually includes features such as a star chart, planet finder and a sort of personal planetarium. More sophisticated software has a variety of additional features. Some of these features can permit you to see the conditions of virtually any predictable astronomical event which involves the relative positions of the sun, moon, planets, and stars. Such events can be current or up to hundreds or thousands of years in the past or future, depending on the software. This can allow you to see the appearance of a conjunction your grandfather told you about, find out if there are any interesting events in the near future, or perform research for various other projects.

The simplicity of using good astronomical software permits you to instantly check out whether or not an event will be visible from your location. For example, while visiting my folks in Arizona, I brought along a computer but did not manage to bring any magazines. One evening, I thought that the moon looked like it could occult Aldebaran in a few hours, but I did not want to stay up in order to find out. Mere minutes at the computer revealed that an occultation would indeed occur from my location; and at what time. This allowed me to go to bed as planned, then just get up for a few minutes to shoot video of the occultation.

Figure 10.9. Left: video frame of the moon a few seconds before it occulted the star Aldebaran on the morning of 27 January 1999. Learning of the event in advance permitted me to capture it on video. Right: astronomical software can be used to predict future astronomical events. This computer simulation with Carina Software's Voyager II program shows the moon and Aldebaran as they appeared at 00:54 on the morning of 27 January 1999, just before the occultation from central Arizona.

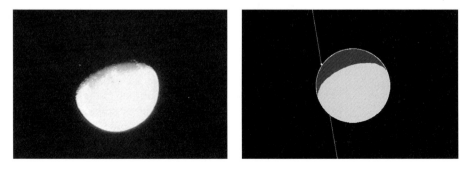

10.10.2 Dating Your Previous Astrophotos

Another application for astronomical software is to research when you took certain conjunction or occultation photos. I typically record the date and time I take various astrophotos, but over the years I have occasionally failed to record data for some photos and misplaced my notes for others. I had all but given up on ever recovering chronological data about these photos before I researched the related events with the use of astronomical software.

Some astronomical programs allow you to find out when past conjunctions or occultations occurred, but most require that you do it in an indirect manner. This typically involves centering the display on a given object, starting an animation at a long time interval, and watching for circumstances which may result in an event like the one you photographed. At this point, you can increase the time resolution and step through the animation to see if the display matches up with your image. This technique can help you to get the data you need, but it can also be time consuming; it took a long evening at the computer just to find the date of two previous conjunctions. Even though it took such a long time to find the dates this way, it was easier than researching a bunch of old periodicals.

Things became a lot easier when Tom Polakis of Phoenix, Arizona turned me on to Carina Software's Voyager II program and I checked it out. Voyager II proved to be one of the more accurate programs, plus it has a feature which puts it head and shoulders above the other programs I tried when it comes to dating astrophotos. This feature is a "conjunction finder" which permits you to select your location, the popular objects of interest, the range of times, and the separation

Figure 10.10. Left: I shot this conjunction photo of the moon (left), Saturn (right center), and Jupiter (lower right) with a 400 mm f/6.3 lens, but I did not record the date and time. The only information I had was the photo lab's date stamp of December 1980 on the slide mount. Right: the same event, simulated with Carina Software's Voyager II program. This program made it easy to determine that the photo was taken at about 7:00 a.m. on 29 December 1980. Of the programs I tried, Voyager II proved to be the best for retroactively dating astrophotos. Its accuracy, combined with a "conjunction finder" feature, allowed me to find the correct date and time for dozens of undocumented conjunction, occultation, and eclipse photos I had taken over the years.

distance threshold. With the click of a mouse, the program calculates all conjunctions falling within the parameters you set, then lists the times and separation distances. You can then view any of the listed events. This feature will also "find" occultations and eclipses with reasonable reliability by just setting a close separation distance. Voyager II allowed me to date 10 years of unrecorded conjunction, occultation, and eclipse photos in a single day!

10.10.3 Practicing Photography of a Future Event in Concert with Computer Software

Astronomical software can also be used to assist you in practicing photography of an eclipse or other astronomical event. All you usually have to do is enter the date, time, and location, then start an animation. (However, some software will not run a "real time" animation unless you set your computer's clock to the time of the event.) This will allow you to preview an event as it will be oriented relative to your surroundings and help you time, practice, and refine your procedures.

If you lack a computer and appropriate software, you can instead practice your photographic procedures in concert with video of a similar event. An even simpler approach is to make some sketches or cardboard cutouts of the event and have a friend display each of them at the appropriate time during your practice run.

Astronomical events won't wait for you to correct errors in your procedures, and when they're over, they're really over! Therefore, practice is important so you can work the bugs out of your gadgets and procedures before the event of interest occurs. By practicing along with an animation, video, or timer, you can get more of a feel for what it will be like to photograph the event itself.

Astrophotography is a rewarding activity in which you can be involved to almost any degree you like. It can be a casual activity you do just for fun or it can be an exacting discipline where you push the limits of your equipment and maybe even invent new gadgets. Just learning about astrophotography can be enjoyable, but the greatest satisfaction typically comes from going out and taking your own pictures. Such pictures are your pictures of the sky as you see it; something you can't get by simply purchasing a poster of someone else's picture.

Experiences associated with astrophotography have rewards of their own. There is nothing quite like being at a dark site with the Milky Way gleaming overhead; the thrill of seeing a bright meteor streak across the sky; the wonder of seeing the intricate structure of nebulae through a telescope; or the familiarity of seeing the moon and planets. Astrophotography does not have to be the emphasis of your outing, but it may help you remember the experiences you had at the time. The best reward of all is simply experiencing the wonders of the night sky.

Appendix A

Astrophotography Exposure Guide

1. Planetary data, including maximum image size @ 20,000 mm focal length.

Table A.1.

Object/Rotation Period at Equator	Mean Distance From Sun (km)/ Orbital Period	Approximate Diameter (km)	Angular Size From Earth (arc seconds*)	Maximum Size (mm) at 20 meters F.L.
Sun 25d	– –	1,392,000	32 arc min.	189
Earth's Moon 27d 7h 43m	149,600,000 27.32d**	3,478	29–34 min.	198
Mercury 59d	57,910,000 87.97d	4,878	4–13	1.3
Venus 243d retro	108,200,000 224.70d	12,104	10–63	6.1
Earth 23h 56m	149,600,000 365.26d	12,756	–	–
Mars 24h 37m	227,940,000 686.98d	6,790	3–25	2.4
Jupiter 9h 50m	778,330,000 11.86y	142,900	30–50	4.8
Saturn (sphere) 10h 14m	1,429,000,000 29.46y	120,600	15–21	2.0
Uranus 17h 14m	2,870,000,000 84.01y	50,800	3.5–4	0.4
Neptune 16h 6m	4,500,000,000 164.79y	48,600	2.2–2.3	0.2
Pluto 6.4d	5,913,500,000 248.59y	2,280	–	–

*   Unless specified otherwise, as in the case of the sun and moon.
** Lunar orbital period around earth shown. As seen from earth, one lunar cycle is about 29.5 days.

Table A.2.

Object	ISO	f/ratio						
	25	–	f/1.4	f/2.8	f/5.6	f/11	f/22	f/45
	100	f/1.4	f/2.8	f/5.6	f/11	f/22	f/45	f/90
	400	f/2.8	f/5.6	f/11	f/22	f/45	f/90	f/180
	1600	f/5.6	f/11	f/22	f/45	f/90	f/180	f/360

Solar system objects	Exposure time in seconds (or as specified)						
Sun, with typical visual solar filter	1/4000	1/1000	1/250	1/60	1/15	1/4	1
Total solar eclipse – chromosphere	–	–	1/2000	1/500	1/125	1/30	1/8
(No solar filter) – prominences	–	1/2000	1/500	1/125	1/30	1/8	1/2
– inner corona	1/2000	1/500	1/125	1/30	1/8	1/2	2
– outer corona	1/125	1/30	1/8	1/2	2	8	30
– extreme corona	1/8	1/4	1	4	15	–	–
Crescent moon – earthshine	1	4	15	1 m	4 m	15 m	–
– <30 hours old	1/125	1/30...	or underexpose twilight sky 1 f/stop				
– 30–60 hours old	1/500	1/125...	or underexpose twilight sky 1 f/stop				
– 20–35% illum.	1/2000	1/500	1/125	1/30	1/8	1/2	2
$\frac{1}{2}$ moon	1/4000	1/1000	1/250	1/60	1/15	1/4	1
Sunlit earth, full moon	–	1/2000	1/500	1/125	1/30	1/8	1/2
Terminator of <50% illum. moon	1/500	1/125	1/30	1/8	1/2	2	8
Moon 4' of arc from terminator	1/2000	1/500	1/125	1/30	1/8	1/2	2
Partial lunar eclipse 0–30%	1/4000	1/1000	1/250	1/60	1/15	1/4	1
(+/– 1 f/stop) 30–80% eclipsed	1/1000	1/250	1/60	1/15	1/4	1	4
(+/– 1 f/stop) 80–95% eclipsed	1/250	1/60	1/15	1/4	1	4	15
(+/– 1 f/stop) 95–99% eclipsed	1/60	1/15	1/4	1	4	15	1 m
Total lunar eclipse (+/– 3 f/stops)	4	15	1 m	4 m	15 m	–	–
Mercury and Venus*	1/2000	1/1000	1/500	1/250	1/60	1/15	1/4
Mars*	1/250	1/125	1/60	1/30	1/8	1/2	2
Jupiter*	1/60	1/30	1/15	1/8	1/2	2	8
Jupiter and four moons*	1/2	1	2	4	15	1 m	4 m
Saturn*	1/15	1/8	1/4	1/2	2	8	30
Uranus (or 6th mag. star)*	4	4	4	4	15	1 m	4 m
Neptune (and Saturn's moon Titan)*	15	15	15	15	1 m	4 m	15 m
Naked eye comet nucleus	6 m	25m...	Typical comet tail = 3× more exposure				

Deep sky objects	Exposure time in minutes (or as specified) (Minimum recommended exposure from dark site)						
Open and globular star clusters	8	30	2h	–	–	–	–
Brighter nebulae, e.g. M42	15	60	4h	–	–	–	–
Most objects, e.g. M20, M51	22	90	–	–	–	–	–
Very dim, e.g. Veil Nebula	30	2h	–	–	–	–	–
12th mag. star (20 cm aperture)	10	Longer exposure required at ~f/16 or slower.					

* The left three columns typically result in a planetary image so small that it lacks significant detail. Data in these columns will overexpose the planetary image in order to make it more visible in the picture. A focal length of at least 5000 mm is required for good detail. To normally expose a small image, extrapolate from the center column by decreasing the exposure by a factor of four in each successive column to its left.

Additional suggestions and data

2. Double exposure if object is only 8–14 degrees above flat horizon.

3. Quadruple (4×) exposure if object is 5–8 degrees high, or for thin clouds.

4. If ISO is higher, decrease exposure time by the same proportion, i.e. if ISO is doubled reduce the exposure time by half. If film is hypered, exposure can be reduced.

5. Keep a record of your own exposure data; it will save film!

6. Use the f/11 exposure data if you are using an f/10 SCT.

7. Extended shutter speed scale (helpful for counting how many stops between speeds)

1/4000	2000	1000	500	250	125	60	30	15	8	4	2
1 s	2 s	4 s	8 s	15 s	30 s	1 m	2 m	4 m	8 m	15 m	30 m
1 hr	2 hr	4 hr	8 hr								

8. Extended f/stop scale

1	1.4	2	2.8	4	5.6	8	11	16	22	32	45
64	90	128	180	256	360	512					

9. Useful focal lengths for various subjects on 35 mm format:

Table A.3.

Subject	Useful focal lengths (mm)
Entire sky	6.0–8.0
Constellations	14–200
Planetary motion, huge nebulae	28–400
Planetary conjunctions	50–2000
Corona at total solar eclipse	200–1500
Sun, moon, lunar occultations	600–2250
Large nebulae (M8, M42, M46)	400–2000
Typical Messier objects (M13, M51)	1000–5000
Small deep sky objects (M57)	2000–10000
Solar and lunar detail	4000–40000
Planetary detail	5000–60000

10. Useful ratios:

1 degree	=	1:57.3
1 arc minute	=	1:3,438
1 arc second	=	1:206,265
1:100 ratio	=	0.573 degrees; or 34.38 minutes; or 34 minutes, 22.7 seconds.

For each 100 mm of focal length, the moon's image will be about 1 mm in diameter when it is at perigee; its image can be more than 10 percent smaller at apogee.

To calculate your f/ratio, divide the focal length by your telescope's aperture.

Appendix B

Exposure Data Form

ASTROPHOTOGRAPHY EXPOSURE DATA - SAMPLE			
SUBJECT / OBSERVATION SITE:			
DRAWING:	**DESCRIPTION:**	**OPTICS:**	**FILM / EXPOSURE:**
Max. Eclipse @ 00:50	Partial Lunar Eclipse 21 Jan. 1981 00:50 local time (MST). From Larkspur Rd., S. of Estes Park, CO.	10.2 cm f/15.5 refractor at prime focus. 1,580 mm F.L. Nikon F camera.	Ektachrome. 200 1/125 sec @ f/15. Same exposure at: 21:39 and 23:50 on 20 Jan; 00:20, 00:30, 00:40 on 21 Jan. 1/60 sec. at 00:51.
	Orion 12 Jan. 1986. Exposure started @ 22:00 MST. From Table Mesa Road, N. of Phoenix, AZ	55 mm f/1.2 Nikkor, stopped down to f/2. Piggyback on C8.	Hypered Ektachrome 400 12 min. @ f/2
	Trifid Nebula (M20) 21 Aug. 1987. Exposure started @ 20:45 MST. From Beaver Creek Rd., N. of Camp Verde, AZ	Questar 3.5 with VersaGuider 3. 1,516 mm F.L. Nikon F camera.	Hypered Konica 1600 110 min. @ f/17
Venus Jupiter	Venus - Jupiter Conj. 23 Feb. 1999 20:10 local time (MST). From Win. Circle in Sun City, AZ.	9.4 cm f/7 refractor with Dakin Barlow & 2x teleconverter. 1,880 mm F.L. Nikon F camera.	Kodak PMZ 1000: 1.5 sec @ f/20. Other exposures: 1/2 sec @ f/20; 1 and 3 sec. @ f/7. Kodachrome 64: 1 sec. @ f/20.

This sample exposure data form shows one way you can record information about your astrophotos as you shoot them. The examples shown here are for some of the photos used in this book. In the order shown here, the corresponding photos appear in Figures 9.4, 5.2, 8.1 and 9.2. A blank copy of the exposure data form is on the next page.

ASTROPHOTOGRAPHY EXPOSURE DATA

SUBJECT:

DRAWING:	DESCRIPTION:	OPTICS:	FILM / EXPOSURE:

Appendix C

Condensed Step By Step Guide to Prime Focus Off-Axis Guided Photography

(See Chapter 8 for more detailed instructions.)

1. Accurately polar align your telescope mount and start its sidereal drive.

2. Locate your subject in a low power (typically 20 mm focal length or longer) eyepiece.

3. Check the appropriate peripheral area around your subject (see Fig. 8.2a) for off-axis guide stars. Also note the appearance of your subject, star patterns, etc., in both your finder scope and camera viewfinder.

4. Attach a guider, reticle eyepiece, camera and counterweight to the telescope. Verify that the reticle eyepiece position will provide a properly focused star image.

5. Point your telescope at a sufficiently bright star within a few degrees of your subject and focus its image in your camera. (Or you can use a focusing attachment.) *Do not adjust your telescope's focus control after you have focused the star image in your camera.*

6. Move the telescope back to your subject.

7. Acquire the guide star you recall from step 3 in the eyepiece on your guider. The object is to rotate the guider on your telescope to a position which will allow its guiding reflector to intercept the guide star you recall from step 3.

8. Loosen the guider's eyepiece holder lock screw and move the reticle eyepiece up or down until the star appears to be in relatively good focus, then lock the eyepiece in place. If you are using an autoguider, attach it at this point, focus it, then go to step 12.

9. Turn on your illuminated reticle eyepiece and verify that the reticle lines look sharp. If they do not look sharp, focus the top part of the eyepiece until they do. Once the lines look sharp, set the illuminated reticle brightness so

the reticle lines are as dim as you can get them without making them hard to see.

10. Center the guide star in the reticle eyepiece, then verify that the reticle lines are roughly parallel with the motion of the star when the telescope is moved in right ascension and declination.

11. With the guide star centered, move your head slightly from side to side while still looking at the star image. If it seems to move side to side in relation to the reticle, adjust the focus by slightly moving your reticle eyepiece up or down until no motion is seen.

12. Rotate the camera to the desired angle for picture composition, accounting for any shadowing by the guiding reflector.

13. Check to see that all lock screws, etc., on your telescope and guider are tight.

14. Attach a locking cable release to your camera, set the speed to "B", and wind the shutter. If your camera has a "T" setting, a cable release may not be required.

15. Be sure you are in a comfortable guiding position, taking into account where the eyepiece will have moved to by the end of your exposure.

16. Precisely center your guide star in the eyepiece reticle.

17. Stand up, stretch out, relax, and check the sky for picture killing airplanes. If you are out alone, get your CD or tape player ready. If you are in a group, show some friends where you will be photographing and see if they will check the area for airplanes and bright satellites every now and then.

18. Precisely center the guide star again if necessary. Take a break to rest your eyes.

19. Open the camera shutter. (Then open your guider's manual shutter if it has one.)

20. Guide your photograph carefully! Look up occasionally to check for airplanes.

21. Close the camera shutter.

Good luck, and may all your star images be round!

Sample Checklist for Star Party or Astrophotography Session at a Domestic Site

This example is based on the setup I used for my 1997 astrophotography outings.

Things To Do At Least One Day Before Outing

* Check out weather prospects for primary and alternate sites.
* Check out condition of car, perform needed maintenance, fill gas tank.
* Test batteries and equipment, charge rechargeable batteries.
* Acquire all necessary film, batteries, video tape, storage media.
* Hyper film (if hypered film will be used).
* Get cash (if needed) from an ATM or other legal source.
* Get all necessary groceries.
* Check and pack equipment.
* Set up equipment and verify that all needed accessories are with it (optional).
* Go to sleep early and/or sleep later than usual.

Things To Do on Same Day As Outing

* Check weather prospects again.
* Check and set clocks or watches.
* Confirm transportation if traveling with another observer.
* Inform family members or other appropriate people that you will return late.

* Prepare water, food, etc.
* Get dry ice if a dry ice cold camera will be used.
* Pack all items that are not already packed; group everything together in one room (if you have space).
* Use checklist to verify everything is packed.

Things To Take On Most Late Out Of Town Trips

Clothing, etc.

* Spare "regular" clothing, particularly if conditions are damp.
* Sweater, jacket and/or coat and hat, as dictated by conditions.
* Blanket or heavy poncho, if and as dictated by conditions.
* Gloves or mittens, if and as dictated by conditions.
* Hat, sunglasses, sun block, provided you will be out during the day.
* Heavy and light shoes, spare socks.
* Plastic bags; for trash and storage of clothes or blankets.
* Optional: umbrella, ear plugs, eye patch, etc.

Food, Hygiene, etc.

* Cold drink bottle and/or thermos, plus at least one gallon of water.
* Food, as required; with all necessary utensils, coolers, etc.
* Facial and bathroom tissue.
* Cup, tooth brush, tooth paste, hair comb or brush.
* First aid, including antiseptic and bandages for cuts or insect bites.
* Medications, as required (asthma inhaler, prescription drugs, etc.)

Documentation and Gadgets not in Equipment Cases

* Emergency cash, telephone calling card, etc., as applicable.
* Wallet, watch, house keys – the stuff you would normally have.
* Davis & Sanford tripod with Aus-Jena head adapter.
* Star D tripod (optional, for star trails).
* Folding chair, camp stool, card table; as required.
* Larger star charts.
* Bright light or lantern (for packing equipment if alone; or for emergency).

Astrophotography Equipment

Telescopes and Accessories in Gray Metal Case

* Vernonscope 94 mm f/7 telescope with tube ring, finder scope, and caps.
* VersAgonal, VersAdapter, telescope adapter, Nikon T-ring, and caps.
* 32 mm Tele-Vue wide field eyepiece.
* Eyepiece rack with 4 mm Meade eyepiece, 8 mm, 12 mm, 16 mm, 24 mm Brandon eyepieces, home made 6× Barlow lens.
* Eyepiece filters in case.
* Amici prism.
* Solar filter.
* Polar alignment guide, astrophotography exposure guide, star and planet finder, Halley finder (to exclude foreground lights), black card, lens paper, lens blower.

Equipment and Accessories In Black Metal Case

* Aus-Jena German equatorial mount head with counterweight shaft, counterweight, heavy brass counterweight knob, counterweight camera bracket, Bogen tilt head, drive corrector, batteries.
* Plastic container with Nikon F body, waist level finder, and cable release.
* Plastic container with Allen wrench, spare hardware, dovetail tripod head adapter, 48 mm Brandon eyepiece, 12 mm Meade illuminated reticle eyepiece with cord, VersAGuider insert 2, rear/top cap for VersAgonal, two inch visual back for Nikon, prism finder for Nikon, spare cable release, spare plastic bags, table cloth.
* Film, video tape, spare 9V & AA batteries, red flash light.

Accessories in Nikon Gadget Bag (optional)

* 35 mm, 50 mm, 105 mm lenses (for piggyback), hot shoe level, Luna-Pro meter.

Telescopes and Accessories in Soft Burgundy Camera Case (optional)

* 300 mm ED Nikkor, VersaScope adapter, DiaGuider, Nikon T-ring, rear cap.

* 20 mm Meade Erfle eyepiece, 9 mm Spectrum orthoscopic eyepiece.
* 67 mm solar filter with 72 mm adapter.
* Slow motion head.
* Lens caps, lens paper, lens blower.
* Nikon 10 × 25 binoculars.
* Canon Photura.

Documentation and Accessories in Nerd Pack

* Road maps; expedition and procedure checklists; checklist for items in cases; books; star chart; exposure guide; pencil, pen, and eraser; writing paper (for recording exposure data); phone numbers; tissue paper; mag light; spare clothes; snacks; prescriptions; water bottle.

Optional Items

* List of planned photos and anticipated exposure times, Gitzo 326 tripod in case with 46 head, Bogen 2025 head, illuminated alarm clock or exposure timer, computer and accessories in case, video camera and accessories in case, audio recorder, party clicker (for occultation timing), portable cassette or CD player and related music or other media, reading material, calculator, thermometer, canned air, masking tape, felt tip marker, small tools, alcohol (to clean some optics), rain tarp, towel, long cardboard dew shield, dew zapper, etc.

Review of Cameras	Lenses	Purpose	Packed in
JVC GR-SZ7	Wide angle attach.	Site video	Black video case
Canon Photura	Built-in	Site photo	Soft case or nerd pack
Nikon F, RA Finder	35, 50, 105 mm	Deep sky photos	Black metal case

Bring additional cameras and lenses for eclipses, meteor showers, other events.

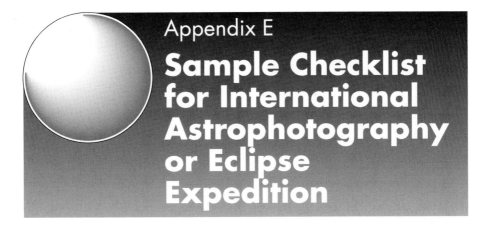

Appendix E

Sample Checklist for International Astrophotography or Eclipse Expedition

This list is based on my 1995 total solar eclipse expedition checklist.

Things To Do Before and During International Expeditions

>3 months before: Determine requirements for trip and equipment; investigate safety.

>2 months before: Book flights, lodging, local transport-ation. Get passport. Get international driving permit and insurance if you will be driving. Contact foreign consulate about medical issues, requirements for temporary importation of cameras, etc. Contact health department regarding required shots; get checkup and any required shots.

>6 weeks before: Set up and test equipment. Make a "do list" if things need correcting.

>4 weeks before: Acquire any additional equipment you may need. Submit documentation of any new equipment to your insurer. Register equipment with customs in your country. Set up and test equipment again. Verify that camera tilt angles required for your subject can be achieved with your setup. Practice packing equipment. Determine and document a final schedule for your trip and photography sessions. Revise

	your "do list". Make a packing list (planning for souvenirs you may acquire).
>1 week before:	Acquire all film, batteries, video tape, etc. Set or check video clocks. Get travelers checks. Determine airport shuttle or taxi schedule.
>3 days before:	Set up and test equipment a final time. Pack equipment. Sleep more than usual, gradually adjust sleep to accommodate time at destination. Ask someone to pick up items left at your door while you are gone.
1 day before:	Set video clocks to WWV. Install new or freshly charged batteries. Pick up and check out any rental equipment you will be using. Stop mail delivery or arrange to have someone pick up any visible mail.
Departure	Turn off utilities. Have passport, tickets, itinerary, and cash accessible.
Upon arrival:	Get bottled water, food, etc.; confirm ground transportation.
Before return:	Confirm your return flight and other transportation.
Day of return:	Try to get boarding passes for all legs of return flight.

Things To Bring on Every International Trip

* Documentation: this checklist, procedure checklists, passport, tickets, itinerary, customs registration certificates, addresses, phone numbers, photos of set up equipment (if necessary for security personnel).
* Financial: body or leg wallet, regular wallet, cash, dimes and quarters, traveler's checks, credit card, calling card.
* Literature: language guide (if needed), star chart(s), exposure guide, books.
* Gadgets: alarm clock, watch, house keys, mag light, pencil, eraser, ball point pen, masking tape, felt tip pen, calculator, blank paper, reading material, international power adapter, two large trash bags, four medium plastic bags, four small plastic bags, luggage with cables and locks (hard carry–on, soft carry–on, plus any checked bags). Laptop computer and cables (optional).
* Clothing: belt, hat, jacket (and/or coat/sweater), shirts, T-shirts, pants, underwear, socks, shoes (and possibly slippers), two pair of sunglasses, photo vest.

* Hygiene: two combs, two tooth brushes, dental floss, tooth paste, mouthwash, Q-tips, shaver with charger and international power adapter, small scissors, soap, shampoo, lip balm, sun screen (>15), facial tissue, toilet paper.
    * Medical: asthma inhaler/nose spray, prescription medicine; aspirin, IBU, diarrhea medication; hydrocortisone, water treatment stuff, insect repellent (with deet if necessary), band aids, small bottles of water and alcohol.
    * Food: peanut butter, soft tortillas, instant coffee, spoon, dull plastic knife, collapsible cup, two filled water bottles.
    * Optional: certified copy of birth certificate, jacket, coat, umbrella, eye patch, cassette player and tapes, shielded muffin fan with cord.

Equipment

Cameras	Lenses	Purpose	Mounting	Film/Tape	Pack In
JVC GR-SZ7	3× converter	Corona video	Star D, slow motion head	JVC ST-C20	Soft carry-on
Sony TR7	0.45×, 0.5×	Umbra video	Bogen C'bar mid right	Sony HG120	Soft carry-on
Canon Photura	Built-in	General	C'bar left, ball	Gold 400	Soft carry-on
Nikon N2020	300 & Barlow	Corona/Visual	Bogen C'bar mid left	K 64	Hard carry-on
Nikon N2020 #5	16 mm	360° panoramas	Panoramic platform	VPS	Hard carry-on
Nikon F, prism	16 mm	First panorama	Panoramic platform	K64	Hard carry-on
Nikon F, prism	14 mm	Umbra, Eclipse	Bogen C'bar right, 3D	Gold 100	Hard carry-on
Nikon F, prism	7.5 mm	All sky	Hand held (for all sky)	K64	Hard carry-on

Accessories	For Use With	Pack In
Film, video tape, tripod case	All cameras	Photo Vest
Soft video case, JVC video case, camera and lens containers	All cameras	Soft carry-on
Batteries, planet finder, rain cover. Luna-Pro meter	All Cameras	Soft carry-on
3× video converter, quick release adapter, 37 mm/67 mm solar filter	JVC	Soft carry-on
0.45× video converter, 0.5× converter, stack adapter, 46 mm adapter	TR7	Soft carry-on
Lenmar video batteries × 4, Sony & JVC battery, charger, copy cables	JVC, TR7	Soft carry-on
Star D tripod, slow motion head; 2.5 inch TV with cable and holder	JVC	Both carry-on

Accessories	For Use With	Pack In
Gitzo 326 tripod, side arm adapter, Bogen crossbar, spare hardware	All cameras except JVC	Hard carry-on
1/4-20 adapter for cross bar, 3D head, 3D head	300 mm, TR7, 14 mm	Hard carry-on
Indexing panoramic platform, vertical camera adapter, hand control	16 mm, 300 mm	Hard carry-on
Cam Lite, Cam Lite with mirror; two cable releases	300 & 2×, 16 mm	Hard carry-on
Hot shoe attachment for Nikon F, flash shoe level attachment	14 mm	Hard carry-on
Cross bar extender, ball head, Ultra Clamp	Photura	Soft carry-on
DiaGuider, Nikon T-ring, 24 mm eyepiece; 67/72 mm solar filter	300 mm and 2×	Soft carry-on
Welder's glass, three eclipse glasses, 10× binoculars, two door peepers	Visual observation	Soft carry-on
Checklist, schedule, maps, eclipse photos, black card, masking tape	General	Soft carry-on
Canned air, lens blower, volt meter, towel or rag	General	Hard carry-on
Tweezers, spanner wrench, four small screwdrivers	Field maintenance	Hard carry-on
Four three-piece posts, >30′ thin yellow barrier tape and/or twine	Barrier for cameras	Soft carry-on
(Optional: Large face digital clock, digital thermometer)	Time and temperature	Hard carry-on

Appendix F
Intellectual Property; Relevant Patents and Allowed Use

1. Recitals

The author and others who contributed ideas for gadgets shown or described in this book desire to retain all commercial rights to their intellectual property, whether or not it is patented. The author has filed and obtained patents for some of his inventions in order to protect his related intellectual property rights; however, he and most contributors also want to allow an individual to freely make and use a device based on some of their respective designs so long as it is for that person's own noncommercial use.

In order to more effectively protect the intellectual property rights of the respective inventors, any other person who makes or uses any of the inventions shown or described in this publication is asked to provide a proper notice about the inventor or any related patents on any copy of the invention said other person may produce or use. In addition, the same people are requested to acknowledge the appropriate original inventor and reference any realted issued or pending patents in any public display or publication which relates to the invention. The purpose of this notification is to help ensure that a commercial interest does not see an unmarked copy of an invention and then proceed to commercialize it without compensating the inventor.

Some inventors of the technology shown or described in this book are seeking licensees for their inventions, so it is possible that others will eventually market authorized products; however, if you obtain such a product, it would be advisable to verify that the vendor is a legitimate licensee who has been duly authorized to make, use, or sell the invention. You can often tell if a vendor is legitimately utilizing another party's patented invention if a relevant patent number is displayed on a related product and their advertising either acknowledges the inventor or indicates that the patented invention is licensed under a specific patent number. Licensing inquiries for the author's inventions and technology may be directed to the Technology Transfer Department at Versacorp (license@versacorp.com).

2. Inventions and Technologies

With the exception of products which are obviously made by commercial telescope, camera, and accessory manufacturers, many astrophotography related inventions shown or described in this book are designed or built by the author. Of these, the author's unpatented inventions, designs, integrations, systems, technology, improvements and trademarks which are shown or described in this publication include:

1.) The embodiment of a Newtonian telescope shown in chapter 2; particularly its mounting, which doubles as a sort of carrying case for the tube and various accessories. Built in 1989. 2.) The particular embodiment of a flip mirror box shown on the back of the Cassegrain telescope in chapter 2. Built in 1983. 3.) The particular embodiment of the axial strut all sky hubcap reflector shown in chapter 2. Built in 1985. 4.) The MicroStar™ focusing attachment described in chapter 2. Authorized products have been offered by the author's company, Versacorp (TM, SM) since 1989. 5.) The three axis adjustable counterweight shown on the 9 cm Cassegrain telescope in chapter 2. Built in 1980. 6.) Serpentine and scalloped spider masks (to eliminate diffraction spikes) described in chapter 4. Built in 1989. 7.) Modifications to the Vernonscope™ 94 mm telescope, Aus-Jena™ mount, and related gadgets shown or described in chapters 2 and 5. Implemented from 1991 to 1994. 8.) The parfocal eyepiece adapter shown on the small Cassegrain telescope in chapter 8. Built in 1987. The authorized product has been offered as a special order item by Versacorp since 1988. 9.) The declination range extender bracket described in Chapter 8. Built in 1987. The authorized product has been offered as a special order item by Versacorp since 1988. 10.) The collapsible dew cap described in Chapter 4. Conceived in 1984. Built in 1986. 11.) The combined Barlow lens and camera lens adapter for the DiaGuider used in taking the "diamond ring" photo on the tripod in Figure 9.9. Built in 1995. The authorized product, the VersaScope™ adapter, is offered by Versacorp. 12.) The indexing rotary panoramic camera platform shown on the tripod in chapter 9. Built in 1991. The authorized product, the Versarama™ is offered by Versacorp. 13.) The vacuum easel and related printing masks shown in chapter 10. Built in 1976.

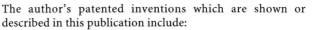

The author's patented inventions which are shown or described in this publication include:

1.) The multiple function VersAgonal™ flip mirror attachment shown on the refracting telescope in chapter 2. Embodiments of this invention are covered by U.S. patents D295,871 and D310,381 and authorized products have been offered by the author's company, Versacorp since 1984. 2.) The DiaGuider™ multiple function image switching attachment and guider shown and described in chapter 2. The DiaGuider is covered by U.S. patent D310,676 and authorized products have been offered by Versacorp since 1987. 3.) The low profile radially adjustable guider shown on the small Cassegrain telescope in chapter 8. The spear prism and eyepiece holder component of this guider is shown in U.S. patent D311,923. The authorized product, the VersaGuider™ has been offered by Versacorp since 1987. 4.) Improvements to the axial strut wide angle all-sky and panoramic capture reflectors described in chapters 2, 9 and 10. The author's original prototype reflectors were not patented, but most of his further improvements of axial strut reflectors (including Cassegrain versions) are covered by U.S. Patent No. D312,263 (Wide Angle Reflector Attachment for a Camera or Similar Article; filed in 1987, issued in 1990) or several dozen utility claims in his pending U.S. and international patents. Licenses and authorized products and technology including the Omnirama™ and OmniLens™ are currently offered only by Versacorp. 5.) The darkroom based panoramic conversion hardware and techniques described in chapter 10 and other forms of light distribution or image processing which are analogous to the same. Implemented in 1976. Patent pending.

Inventions by others which are shown or described in this publication include:

1.) David Charles' enhancements and attachments for the "Barn Door" mount shown and described in chapter 6. Built in 1986. 2.) Pierre-y Schwaar's "Picaddilly" adjustable partial aperture guider which is mentioned in chapter 8. Built in 1986. 3.) Telescopes, cameras, accessories, and other commercial products by various manufacturers. Any relevant patent numbers are usually on the subject products.

3. Conditions of Use

Allowed use of the disclosed intellectual property. The author and/or some contributors have invested considerable time and resources in developing their intellectual property and reserve all commercial rights to all aspects and accounts of their inventions, designs, integrations, systems, technology, discoveries; and improvements or reductions to practice of any items, methods, techniques, literary works, or visual art which they each originated or uniquely improved.

At the present time, the author typically grants an individual a free license to build one of his amateur optical or accessory items which relate to telescopes and reflective wide angle optics (not a copy of a commercial item, but instead an item like one of the amateur systems shown and described in this book or one of the author's previous [1998 and earlier] amateur papers) for their own private non-commercial and nonprofit use (i.e. make and use for own noncommercial purposes, but not sell them or use them for profit without paying applicable royalties) provided that they affix a readable label to the item which includes the statement: "U.S. Patent [any relevant patent number goes here*] and Patent Pending [in cases where a patent is pending]. Used by permission of Jeffrey R. Charles." and specify the same in any publications or public presentations relating to the subject systems. All commercial, contract, and publication rights are reserved. Any other use without the prior express written consent of Jeffrey R. Charles and the payment of applicable royalty is strictly prohibited.

The patent notice is not necessary for a wide angle refelctor if the constructed embodiment is exactly like one of Jeffrey R. Charles' pre-1987 prototypes which were the subject of publications up to and including August, 1986, since those particular two prototypes are not patented However, it would be appreciated if people who build a copy of any of these prototype units acknowledge Mr. Charles as the inventor.

The author typically enforces his patents in regard to commercial vendors, users, applications, government contracts, etc.; and reserves the right to do so in other cases where a substantial quantity of units is made, sold, or used without authorization and the payment of prescribed reasonable royalties. All rights (including commercial and intellectual property rights) reserved.

Thank you for your indulgence.

The use policies herein are subject to change without notice, and their scope may not be the same in all countries, depending on local intellectual property laws. The author (and any contributors holding applicable patents) reserve the right to require prior approval and a nominal royalty even for each individual use in any country where allowing free individual use of an invention may restrict commercial enforcement of any applicable patent(s). The author, contributors, Versacorp, and any licensees who obtain exclusive rights to technology belonging to any of any of the same reserve the right to stop granting licenses for the invention(s) involved to individuals, nonprofit groups, or others with or without notice at any time.

---

* The appropriate patent number(s) can be obtained from the list of patented inventions in this section or the bibliography or some of the author's other works. The part of the notice which contains patent information should also indicate if patent(s) are pending. For example, a patent notice for an appropriately improved version of the author's axial strut wide angle reflector would read: "U.S. Patent D312,263 and patents pending..."

Appendix G

Bibliography and Other References

References in Print by Other Authors

Charles, David, L., "Serious Astrophotography with the Barn Door Mount", in Proceedings of the 1988 Riverside Telescope Makers Conference, OCA Publications, pp. 60–67. Covers the clever and economical implementation of a tangent correction cam and other enhancements for a "barn door" tracking mount.

Covington, Michael, Astrophotography for the Amateur, pp. 61–66. Calculations for Barlow lenses and other auxiliary optics.

Haig, G. Y., Sky and Telescope, April 1975, p. 263. Early and informative article by Haig about what is now commonly called a "barn door" mount.

Lucien Rudaux and G. de Vaucouleurs, Larousse Encyclopedia of Astronomy, Prometheus, 1959, p. 400. Background for Messier catalog of nebulae and clusters.

Malacara, Daniel (ed.), Optical Shop Testing, John Wiley, 1977, pp. 351–355. Star tests, characteristics of the Airy pattern.

Menzel, Donald H., A Field Guide to the Stars and Planets, Houghton Mifflin, 1964, p. 381. Table 18, Planetary Data, includes planet size and orbit data. Also notable because it includes photographic star charts of the entire sky.

Questar 1977 Catalog, p. 24. Illustration of Airy disk and diffraction rings.

Reynard Corporation 1992 optical components catalog, p. 74. Table of "percent reflectance at normal incidence of freshly evaporated mirror coatings".

Rutten, Harrie and van Venrooij, Martin, Telescope Optics: Evaluation and Design, Willman-Bell, 1988, p. 208. Illustration of the Rayleigh Criterion.

Shafer, Rick, Your Guide to the Sky, Lowell House, 1993, p. 160. Solar system objects comparison table.

Sidgwick, J. B., Amateur Astronomer's Handbook, third edition, Faber and Faber, 1971, p. 116. Characteristics of rhodium mirror coatings.

Sky and Telescope, August 1986, p. 186. Coverage of 1986 Riverside Telescope Makers' Conference. Includes photos of Jeffrey R. Charles's early axial strut wide angle reflectors and David L. Charles's enhanced barn door mount.

Publications on the Internet by Other Authors

The URLs of many web sites seem to change often, so a printed list can quickly become outdated. Accordingly, web sites by other authors are not listed here. Relatively up to date links to appropriate astronomy related sites are usually listed in the "Links to Material by Other Authors" section of my Eclipse Chaser (www.eclipsechaser.com) or Versacorp (www.versacorp.com) web sites. Other links and articles can usually be found at the web sites of popular astronomy related publications and agencies.

References by Other Authors on Computer Readable Media

Voyager II Dynamic Sky Simulator, by:

Carina Software
12919 Alcosta Blvd. Suite #7
San Ramon, CA 94583
(925) 355-1266 www.carinasoft.com

This software made it relatively easy to retroactively date some of the astrophotos in this book.

References in Print by Jeffrey R. Charles

"Adventures in Astrophotography with a Small Telescope", in Proceedings of the 1988 Riverside Telescope Makers' Conference, OCA Publications, pp. 83–91. Includes a step by step guide to off-axis guided photography which was the basis for the one in this book.

"Innovative Space Mission Applications of Thin Films and Fabrics", in Proceedings of the 1995 Jet Propulsion

Laboratory miniconference, JPL, pp. 61–82. Airborne Large Aperture Telescope (ALAT).

"Portable All-Sky Reflector with 'Invisible' Camera Support", in Proceedings of the 1988 Riverside Telescope Makers' Conference, OCA Publications, pp. 74–80. Shows and describes standard and Cassegrain versions of my axial strut wide angle reflectors, suggests improvements, and addresses the use of reflectors for both all sky and one shot 360 degree panoramic imaging.

Instruction manuals for the patented Versacorp DiaGuider (a combined flip mirror and guider); MicroStar (a focusing attachment); VersAdapter (a multiple function eyepiece holder); VersAgonal (a multiple function flip mirror); and VersaGuider (a series of adjustable guiders and guider inserts).

US Patents D295,871, D310,381, D310,676, and D319,450 for Image Switching Attachments (flip mirrors); D311,923 for a Telescope Guider; D312,087 for a Telescope Converter Attachment; D312,263 for a Wide Angle Reflector; PCT patnt application for an Omniramic™ (360 degree panoramic) optical system; and, PCT patent application for an Omnidirectional (full sphere) Optical System.

With Robert Reeves and Chris Schur, "How to Build and Use an All-Sky Camera", edited by Richard Berry, Astronomy, April 1987, pp. 64–70. Jeffrey R. Charles was the sole author of "The Strutless All-Sky Camera" section on pp. 68 and 69, which describes one version of his axial strut wide angle reflector.

References on the Internet by Jeffrey R. Charles

Eclipse Chaser (www.eclipsechaser.com) web site. An award winning resource for astronomy, eclipse, and wide angle imaging enthusiasts. Includes numerous technical articles, eclipse expedition journals, editorials, images, and even a novel and screenplay.

Versacorp (www.versacorp.com) web site. Includes product information and technical papers. Originated June, 1996. Also linked to the Eclipse Chaser web site.

Index

Index